D1005944

DESIGNER ANIMALS:
MAPPING THE ISSUES IN ANIMAL
BIOTECHNOLOGY

Designer Animals

Mapping the Issues in Animal Biotechnology

Edited by Conrad G. Brunk and Sarah Hartley

UNIVERSITY OF TORONTO PRESS
Toronto Buffalo London

Toronto Buffalo London
www.utppublishing.com
Printed in Canada

ISBN 978-1-4426-3997-3

Printed on acid-free, 100% post-consumer recycled paper with vegetable-based inks.

Library and Archives Canada Cataloguing in Publication

Designer animals : mapping the issues in animal biotechnology / edited by Conrad G. Brunk and Sarah Hartley.

Includes bibliographical references and index.
ISBN 978-1-4426-3997-3

1. Animal biotechnology – Social aspects. 2. Animal biotechnology – Moral and ethical aspects. 3. Animal biotechnology – Economic aspects. I. Brunk, Conrad G. (Conrad Grebel), 1945– II. Hartley, Sarah Annette

SF140.B54D48 2012 179'.3 C2011-906917-2

University of Toronto Press acknowledges the financial assistance to its publishing program of the Canada Council for the Arts and the Ontario Arts Council.

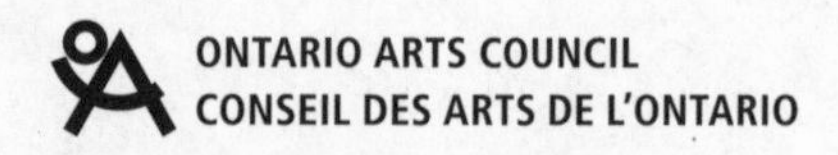

University of Toronto Press acknowledges the financial support of the Government of Canada through the Canada Book Fund for its publishing activities.

To Christiane; and to Felix and Abigail, whose births during this project gave us hope for the future

Contents

List of Figures and Tables

Acknowledgments

This book is the product of an intensely collaborative effort by a large team of people without whom it could not have succeeded. The editors had the privilege of coordinating the efforts and expertise of an extraordinary group of researchers, authors, advisers, and supporting staff. This made our job both effortless and pleasurable.

Our research project and this book were shaped by a research model developed by Harold Coward, one of the contributing authors. The model provides a way of conducting a broadly interdisciplinary research enterprise in a way that creates a 'common mind' among the contributors, so that the resulting chapters of the book reflect a coherent discussion and debate of an issue, rather a mere collection of individual viewpoints on the issue. This is achieved by bringing together all the authors, research assistants, advisers, and support staff in two separate meetings – the first to discuss the unifying theme of the book and the approach of each chapter, and the second to review and critique first drafts of the chapters. Each author's final contribution incorporates the feedback from all the other members of the team. This process requires a far greater commitment of time, travel, and collaborative energy than the usual stand-alone, disciplinary research project with which most academic authors are familiar. The editors wish to thank all the authors for their enthusiastic commitment to this process.

The methodology of the project involved the use of a series of focus groups with persons who identified themselves as having a particular 'stake' in the advancement of animal biotechnology – that is, they were involved in, or impacted by, the technology by virtue of their professional or personal roles and commitments. We are indebted to all those focus group participants who were willing to take the time out of their schedules to discuss these issues. The focus groups were organized in various locations across Canada and the United States through the tireless efforts of two research assistants, Paul Teel and Jean-François Sénéchal. They were expertly facilitated by Leslie Rodgers of Praxis Pacific, whose commitment of time, travel, and background research far exceeded the terms of her contract. The rapport she so evidently established with the participants was key to establishing the free and creative debates on sensitive issues that characterized these groups.

The scientific, philosophical, and regulatory integrity of this project was greatly enhanced by the expertise of a group of advisers who participated in team meetings and in the review and critique of chapter drafts. This group included David Fraser, Cecil Forsberg, Keith Pitts, Michelle Illing, and Amanda Whitfield.

The editors are indebted to the director, Paul Bramadat, and the staff of the University of Victoria Centre for Studies in Religion and Society, who managed all the administrative details of the project, including organization of the team meetings and management of the many chapter drafts. Leslie Kenny, the centre administrator, expertly supervised the work of Rina Langford-Kimmett and Susan Karim. In addition, she coordinated the financial details of the project and prepared the manuscript. Leslie probably devoted more time and energy to this book than anyone, and it could not have been brought to fruition without her excellent work.

This project was funded in its entirety by a Canadian Network of Centres of Excellence grant through the Advanced Food and Materials Network (AFMNet). The collaboration of the editors, authors, and research assistants with other scientific and GE^3LS (genomics, ethics, environment, economics, law, society) projects in the AFMNet, coordinated by its scientific director, Rickey Yada, added considerable value to this project.

DESIGNER ANIMALS:
MAPPING THE ISSUES IN ANIMAL
BIOTECHNOLOGY

1 Introduction: Focusing on the Values in Debates about Animal Biotechnology

CONRAD G. BRUNK, SARAH HARTLEY, AND LESLIE C. RODGERS

Since the initial discovery of the double helix, genetic engineering by means of recombinant DNA technologies has inspired the imagination of those who envisioned its many potential benefits. It has inspired hope for a new 'green revolution' that would significantly increase agricultural productivity around the world, enhance the nutritional quality of food, allow the development of new therapeutics, and permit the control if not elimination of genetic diseases. In the minds of some, it promised also the improvement of the human species through the enhancement of genetic-based characteristics in future generations.

But it also sobered the imagination of those who envisioned the potential risks and abuses of the technology. The worries ranged across a wide spectrum of possible consequences, including the potential for detrimental impacts of new organisms introduced into ecosystems, adverse health effects of genetically engineered microorganisms or plant and animal food products, and also ethical concerns about 'playing God' with plant, animal, and, especially, human nature. Many of the ethical concerns about genetic engineering have focused on its application to humans, with issues around cloning and eugenic biotechnology taking the forefront in the ethical debate (for example, Habermas 2003; Fukuyama 2002; McKibben 2003). These concerns have been articulated largely in terms of impacts upon human dignity, autonomy, and well-being.

The debate on the questions of values and ethics posed by animal biotechnology emerged in the 1990s in response to scientific speculation about the types of products and applications that might be derived from cloned or engineered animals. This debate remained limited and was confined to the academic literature (Holland and Johnson 1998; Rollin 1995; Thompson 1997). The scholars who did take up the debate highlighted a range of ethical issues, including those of environmental and human health, the 'integrity' of individual animals and animal species, and animal rights and welfare, among others. This debate was based largely on speculation about the potential uses of cloned and engineered animals, well before most of the laboratory successes with the technology that are now beginning to emerge.

It should not be surprising that the ethical concerns about genetic engineering and biotechnology (the terms are used interchangeably in this book) with respect to human beings would also find their parallels in the context of plant and animal biotechnology. The public, as well as the scholarly, debate about plant and animal biotechnology was not limited solely to the empirical questions about adverse impacts upon human health and the environment, but also raised questions about the ethics of producing organisms of certain types that seem to stretch the limits of the 'natural.'

Some people worried that the very transfer of genes from one species to another, a process that would not happen otherwise via 'natural' processes, was an unacceptable intervention in nature – a concern often expressed in the vague, but powerful, terms of 'playing God.' The broader impulse underlying the 'playing God' objection, shared even by those who would not in other contexts invoke theological language, seems to be a feeling of sacrilege – something that violates a sacred value. If one holds that nature is in some sense 'sacred,' then, as with all things sacred, it is to be revered and approached with a sense of awe. The sacred is not to be touched, except in ways that respect its sacred character. This impulse is always difficult to articulate, and thus the 'playing God' concern is often downplayed by philosophers and those with more pragmatic interests in the technology. It may have something to do with the sense of 'abomination' people feel when confronted with phenomena that do not fit within their normal ways of conceptualizing the world in which they live.[1] The relevance of this idea will be discussed in later chapters of this book.

People have been crossing species of plants, and even some animals (for example, the mule), by conventional breeding and hybridization techniques for centuries. Hence, broccoflowers, tangelos, nectarines, and other cross-species fruits and vegetables raise very little, if any, sense of sacrilege or abomination. But when people are confronted with the prospect of fruits and vegetables engineered with genes taken from fish, pigs, or other animals, the so-called yuk factor begins to emerge, sometimes to the point of a sense of abomination or sacrilege. The same response is produced in many people at the prospect of animal chimeras (e.g., a cross of a cat and a rabbit, a pig and a sheep), and in even more people at the prospect of a human/non-human animal chimera. Further, the higher the animal in the phylo-genetic scale, and the closer the bonds of companionship people form with animals, the more likely they are to invoke the ethical considerations invoked in the case of humans themselves – considerations of well-being, dignity (or 'integrity'), and even autonomy.

Nearly everyone, it seems, feels that there are limits to how biotechnology should be used, but most find it hard to articulate precisely why they draw the lines of limitation at the places they do. Most recognize that the values involved in these questions are many and complex, and different people weigh these values and trade them off in different ways. This book is an attempt to take a careful look at how people with various stakes and interests in animal biotechnology tend to identify the fundamental values – moral and non-moral – in this field and how they weigh them in their thinking about how biotechnology should or should not develop.

This book represents a collective effort to understand the range of stakeholder concerns about animal biotechnology, and how these concerns should be managed or regulated as animal biotechnology moves from the laboratory to the marketplace. While public opinion studies show that animal biotechnology triggers more ethical and value concerns than other areas of biotechnology, such as bio-medical or plant biotechnology, there has been virtually no empirical research on and little theoretical discussion of these issues associated with specific applications of animal biotechnology or of the ethical and value concerns among a plurality of stakeholders. Two exceptions to this include the excellent public opinion assessment conducted by the Pew Foundation in the United States (Pew Initiative on Food and Biotechnology 2006) and the Danish study Cloning in Public (Gamborg et al. 2006).

In the last decade, a small number of public opinion polls commissioned to test consumer or citizen reaction to animal biotechnology joined the academic debate (Pew Initiative 2006; Pollara and Earnscliffe 2003; Saad 2004). In general, these empirical studies found that the general public has a strong aversion to animal biotechnology, although they provide little understanding of the reasons for such an aversion.

For example, a Gallup poll found that 64 per cent of Americans considered animal cloning morally unacceptable (Saad 2004). A Pollara and Earnscliffe (2003) poll found that while 64 per cent of Canadians supported animal cloning for medical research, only 24 per cent supported cloned animals in the food system, and, interestingly, it was the ethical issues that drove the public's perception of acceptability in the area of animal biotechnology. In a more recent poll, Co-op, a British food retailer, contacted over 100,000 consumers to better understand public concerns when developing a responsible retailing policy. The supermarket chain found that 21 per cent of consumers rated animal welfare as the top ethical priority, well above fair trade (14 per cent) and climate change (4 per cent) (Guardian Unlimited 2008).

Most opinion polls and other empirical research done on public attitudes towards biotechnology do not take into account the impact that one's 'stake' in the issue can have upon one's attitudes. At most, poll results will be broken down in terms of age, gender, and perhaps socio-economic status. However, when people are asked to give their opinion in response to a specific question, their answer can be significantly influenced by their assumptions about the *appropriate* answer. What they consider *appropriate* may depend upon what they are assuming about their social *role*. When asked a question out of any specific social context, most people are likely to answer it in terms of their personal ideals and values, or perhaps in terms of their ideal of responsible citizenship. One of the uncertainties of much opinion research is that it will not be known how these assumptions about role are influencing responses.

This book attempts to characterize the debate about animal biotechnology in a way that takes seriously the ways in which people's attitudes are influenced by the interests, or what might be called the 'stakes,' they have in the technology. As with any technology, some people stand to benefit from it far more, or in different ways, than others; and some people may see their interests more likely to be threatened, or placed at risk by it, than others. So, depending upon one's social position relative to the technology, different values – different goods to be gained or placed at risk – will be more salient and carry more weight.

So, we have engaged authors to write the chapters of this book from the point of view of how animal biotechnology – the genetic engineering and the cloning of animals – promises to serve different kinds of social interests and raise challenges for certain moral, religious, and other social values. We identified seven social and economic sectors that we felt were most involved in this technology, as producers, as deliverers, as direct or indirect beneficiaries, as risk bearers or representatives of those at risk (e.g., patients, the animals themselves), or as persons whose personal or social values were promoted or challenged. This identification was determined in large part by the direction the technology development has taken, and the major implementations that have been envisioned. We wanted to know how people who had a primary professional or social identification with one of these sectors, and who thus in an important sense might share values common to that sector, would view the value questions posed by the technology. It is in this sense that we call these people 'stakeholders': their professional or social role identifies a set of 'stakes' they have in the debate. 'Stakes' are the values they have a special social role in promoting. To call them such is not to judge their merits in any way. We prefer not to call them 'biases' (a pejorative term).

The study utilizes two complementary methods for assessing the way in which the animal biotechnology debate takes shape within North American society. The first is the largely theoretical exercise of reviewing how the technology is developing in the range of selected sectors, identifying the interests and values driving that development, and assessing the value conflicts and challenges most likely to arise within each sector. We have asked the various chapter authors to do a careful review of the state of the technology in each sector, and of the literature assessing the benefits and promises, as well as the risks and the challenges.

We have, however, complemented this theoretical exercise with the employment of focus groups that included well-informed actors within each of the sectors. We identified the following seven sectors as having major stakes in animal biotechnology:

1 The Scientists, who provide the basic scientific research that facilitates the development of cloned and genetically modified animals in the laboratory. This focus group was recruited at the UC Davis Animals in Biotechnology Conference (Tahoe City, CA) – an international conference of leading scientists in the field.
2 The Agricultural Producers (farmers, processors, retailers) of cloned or genetically engineered animals, who develop new products and bring them to market. This focus group recruited people who were

leaders in these fields, in most cases involved in organizations representing these interests.

3 The Health Researchers, including funding agencies, who develop new health therapies based on animal biotechnology. This focus group recruited scientists working in health research in universities and other research institutions.
4 The Health Care Providers and Patient Advocates, who use the therapies provided by animal biotechnology. This focus group was composed of nurses and active advocates of patient interests in Canada.
5 The Animal Justice Advocates, who claim to represent the interests of the animals themselves. This focus group recruited leaders in influential animal rights and animal welfare organizations.
6 The Promoters of Alternative Agriculture, who represent the concerns of social justice in farm communities and in global development, including organic farmers. This focus group included people active in organic farming, farmers' markets, and other alternative agriculture organizations.
7 The Government Regulators, who have the responsibility to approve the introduction of new technologies into the market. This focus group was recruited from Canadian regulatory agencies with a statutory responsibility for regulating the products of animal biotechnology.

Other stakeholders arguably could also have been included in this study. Pet owners and the users of animals for sport and recreation come immediately to mind, given the potential value of biotechnology to clone deceased companion animals or to enhance desired performance characteristics through genetic engineering. It was necessary, for obvious reasons, to limit the scope of the study. However, because of the important role that religious world views and values play in the formation of many people's opinions on animal biotechnology, we included a chapter devoted to this issue. Because of the wide variety of religions and religious perspectives represented in our pluralistic society, it was not feasible to include a focus group comprising people who could be asked to view the issues around animal biotechnology from the perspective of their religious values. In lieu of a religious focus group, we engaged the other stakeholder groups in debate about possible religious values at play.

The aim of this much more directly empirical methodology was *not* to provide a reliable sample of opinion representative of the stakeholders within the sector. Focus groups can rarely provide a reliably representative sample. The value of the focus group research in this project is to engage in the debate around animal biotechnology well-informed major actors/stakeholders in order to identify nuances in the perspectives of people with such stakeholder involvement that might otherwise go unnoticed. We asked the authors of the chapters to analyse carefully the transcripts of the focus group discussions in order to enrich their own understanding of the way the value questions are shaped by some of the stakeholders. The comments made in the focus groups are in no way represented in these chapters as evidence in themselves for 'the way things are seen' from that stakeholder perspective. Instead, the comments made by these informed participants are used as examples of commonly held viewpoints, or, perhaps more importantly, as indications of sensitivities that may be thought to be absent among persons with the particular stakeholder interest (for example, agricultural producers are often accused of being uninterested in animal welfare unless it contributes to higher yields).

Despite responding to questions 'in the moment,' our focus group participants often articulated their views with eloquence and clarity. As is typical in focus groups, some participants spoke more often than others, or had a particular knack for expressing a view that resonated with others in the group. Hence participants are quoted with varying frequency. Overall, we are confident that quotes have been used judiciously and reflect the tenor of the respective groups.

Our agreement to maintain anonymity prevents us from identifying individual participants or even organization names. However, we have assigned an indicator (e.g., V1, V2) for members of respective focus groups to connect the person to quotes in the chapters.

An additional comment on the methodology of this book: In order to achieve maximum coherence and integration in the book despite its multiple authorship, we engaged all the authors in a process of interdisciplinary discussion and mutual critique in the book's production. The authors have all participated in a team process that included two three-day workshops. In the first workshop, the general perspective and unifying theme of the book was shaped and the structure of each chapter laid out. The authors then went their ways to do the research and writing of a draft chapter, which was then brought back to a second workshop where each chapter author was given a critique by all the others. At this second workshop, the whole project team discussed the transcripts of the focus groups and the means of integrating them into the chapters. The authors then finalized their chapters in light of this process.

The term 'biotechnology' is used in many different ways in different contexts and for various purposes. So it is important to clarify that, for the purposes of this study, the term 'animal biotechnology' is used in a way that is somewhat restrictive in one sense, and very broad in another. The term is often used to refer to any technology by which the genome of an animal is intentionally reshaped to produce novel phenotypic characteristics. Used in this sense, the term embraces conventional selective breeding, hybridization, and mutagenesis techniques (in plants and animals) as well as direct manipulation of specific genes through such procedures as recombinant DNA, 'gene knockout,' and so on. In this study, the term will refer only to the latter technology – the direct manipulation of genes in the animal genome to produce genomic variations that could not be achieved by other conventional means.

However, the term 'animal biotechnology' will also be used more broadly to include techniques for the reproductive cloning of animals. Technically, reproductive cloning, or somatic cell nuclear transfer (SCNT), does not itself involve the intentional manipulation of the genome, since its aim is to reproduce an animal that is genetically as similar as possible to its progenitor. SCNT is included in this study as an aspect of animal biotechnology because of its controversial nature as a reproductive technology in human and non-human animals, and because it is the biotechnology that currently is most developed for introduction into the market (indeed, cloned animals and their progeny are now available on the American market). For our focus group participants, we defined animal biotechnology as 'the range of scientific tools and techniques involved in the deliberate manipulation of DNA in animals (genetically modified, or GM, animals) and cloned animals.'

The Focus Group Research

The focus group research that underlies this study consisted of seven focus groups conducted in Canada and the United States between August and December 2007. A professional moderator facilitated these focus groups, assisted with the research design, and transcribed and coded the focus group discussions. We designed the focus groups to illuminate the values and ethical concerns that underlie people's views on animal biotechnology, as seen through the lens of their particular stake in the subject. We recognized that the questions were likely to challenge participants to think about and articulate those values at a depth they may not have previously explored. For that reason, we felt the participants would be more comfortable discussing the issues if the focus groups consisted of people with relatively homogeneous interests, rather than heterogeneous groups that might result in positional, ideological, or debate-style dialogue.

The research team identified potential persons to invite to the focus groups by investigating membership lists of organizations associated with the stakes in animal biotechnology we wished to include in the study. We also identified individuals who were active in public debate around agriculture and biotechnology or were associated with one of the seven targeted sectors. Since the pool of informed stakeholders in Canada from which to draw was relatively small, participant recruitment for the majority of focus groups required multiple efforts and considerable follow-up. However, when compared to typical response rates for focus groups, our rate of response in most areas suggested significant interest in the subject. Our focus group participants confirmed this level of interest:

> I really think that it is important to have alternative perspectives expressed in the literature and have more knowledge about those alternative perspectives. (Alternative Agriculture stakeholder V1)

> When a person gets the opportunity to participate in something like this on the front end, I don't like to not take the opportunity because we see a lot of things that come out on the back end and you go, 'I sure wish I would've had a chance to speak in the formative stage of that piece.' (Agricultural Producers stakeholder V5)

> It's become obvious to me that ethics has come up as . . . the major hurdle that this technology has to get over in order to get to the market, and too few scientists are willing to discuss the ethical issues, and I think we have a responsibility to get our own voices into this discussion and not let it be driven by people with other agendas. (Scientist V6)

> The ethical debate is something we can't ignore . . . When given the opportunity to participate in it, I think we should take advantage of it. (Regulator V3)

Our goal was to engage a minimum of four and a maximum of eight participants per focus group, as we believed groups of this size would reveal various views yet allow for deep, values-based discussion. As table 1.1 shows, we achieved our goal in all but one group.

The moderator followed the same format with each of the focus groups, beginning with a few introductory questions that explored participants' knowledge of existing animal biotechnology applications as well as the hopes and possible benefits of the technology and the risks and concerns. The moderator carefully designed the line of questioning to move from general to specific, from familiar to less familiar, and from top-of-mind to in-depth as participants' comfort increased with the process, the moderator, and one another.

The substance of the focus groups was the discussion that followed the ranking of ten carefully selected sample applications of animal biotechnology. We provided participants with ten actual or theoretical animal biotechnology applications, selected to flush out stakeholders' values and trade-offs based on the type of application (food, human health, environmental, industrial, sport/commercial), the animal involved (whether of a 'lower' or 'higher' order in the animal kingdom), the process, the product, and the potential beneficiaries of the application. Since we were particularly interested in where people might draw the line between what they perceive as acceptable and unacceptable, we included a very broad range of possible applications. The ten applications included:

1 Cloning of a prime breeding bull to enhance production of high-quality progeny (e.g., increase milk production)
2 Cloning to help preserve or revive animal species that are endangered or extinct
3 Genetically modifying a pig to produce omega-3 fatty acids in its meat to benefit human health
4 Genetically modifying a salmon to make it grow faster and feed more efficiently for aquaculture production
5 Genetically modifying a cow to produce human insulin in its milk for treatment of diabetes
6 Genetically modifying a laboratory mouse to increase its susceptibility to cancer, making it more suitable for cancer research
7 Genetically modifying a primate to grow a human brain for tissue transplantation to brain-injured humans

Table 1.1
Summary of focus group recruitment and participation

Group	Recruitment pool	Confirmed	Attended
Scientists	85	7	8
Regulators	12	8	7
Alternative Agriculture	8	5	4
Animal Justice	9	4	3
Agriculture Producers	9	6	6
Health Care Providers and Patient Advocates	20	6	5
Health Researchers	40	5	5
	T = 183	T = 41	T = 38

8 Genetically modifying a pig to reduce the amount of polluting phosphorous in its manure
9 Genetically modifying an aquarium fish so it can re-emit light, giving the appearance that it glows
10 Genetically modifying a goat to produce spider silk in its milk for industrial production

The moderator asked participants to rank the applications from the most supportable to the least supportable, and then probed the participants to tease out the values underlying their choices and the nuances between conflicting values. The moderator encouraged participants to examine the reasons for their largely visceral reactions, the range of moral and other values informing their choices, their 'tipping point' (where they drew the line between supportable and unsupportable applications), and how they traded off benefits and concerns.

Following the ranking exercise, the moderator probed the focus group participants on questions of governance. These included who should be involved in governance, current levels of confidence in safety and regulation, how government should consider and incorporate public values, and what people would want to know about the animal products resulting from animal biotechnology.

Focus group participants identified the full range of ethical issues associated with animal biotechnology as identified in the academic literature and public opinion polling. This suggests that the combined groups reached 'saturation,' that is, the point where the range of ideas and perspectives was likely heard and additional groups would not likely have yielded substantially new information. The top-of-mind benefits participants named focused largely on human health (new, less costly, and more-effective diagnostics and treatments); increased productivity, supply, and quality of food animals (to 'feed the world' and to improve nutrition, taste, variety, cost, and the safety of animal protein); and environmental benefits (animals that produce fewer pollutants, utilize food and nutrients more efficiently, or neutralize pollutants). Other benefits cited included economic ones (for producers of animal biotechnology products and the consumers purchasing those products), animal health and survival (improved disease resistance, higher-quality milk for offspring, cloning to preserve endangered species), and a general contribution to scientific knowledge.

Some stark differences emerged between focus groups with respect to hopes for animal biotechnology. A hope expressed in the Scientists, Regulators, and Agricultural Producers focus group was that the public would accept animal biotechnology:

> I hope that the unimaginable is out there, whether it's a cure for something that we've been looking for, or a better foodstuff, or who knows what. And I hope that society will let us pursue animal biotechnology to be able to find that. (Agricultural Producers stakeholder V5)

On the other hand, participants in the Animal Justice and Alternative Agriculture focus groups more generally hoped, respectively, for a complete or a virtual halt to the technology:

> My only hope around biotechnology food products is that it doesn't happen. It's eliminated. (Alternative Agriculture stakeholder V1)

Participants identified a considerable range of risks and concerns related to actual and potential impacts on animals – physical, emotional, and social – as well as on individual animal and species 'integrity.' For example, in the Animal Justice group, any human-inflicted animal pain and suffering was seen as unconscionable:

> So, in my view, if an animal is capable of feeling pain and suffering, that creates the obligation on us not to cause that pain and suffering, even in

> the name of ending some other pain and suffering which may be an evil, of course, but it's not an evil that that animal is in any way responsible for. (Animal Justice stakeholder V2)

In other groups, participants tended to wrestle more with the notion of animal pain and suffering that might be unfortunate but 'necessary' to achieve a broader societal good (usually a significant health benefit for a significant number of people), versus that which is 'unnecessary' (because, for example, alternative approaches are available, or because there are too few benefits or beneficiaries to justify the cost).

Other major risks and concerns included moral and ethical issues (such as 'playing God,' unnaturalness, lack of respect for life, profit motivation, and biotechnology being a 'slippery slope'); risks to humans and the environment (due to unintended or unknown consequences, loss of genetic diversity, and intermingling of genetically modified and wild animal populations); and economic and social impacts (to the developing world, to organic and conventional meat producers, and to Canadian meat exports).

Most of the two hours allocated for each focus group was spent discussing the sample animal biotechnology applications. Participants in all focus groups ranked three of the sample applications 'most supportable' considerably more often than the others – the lab mouse genetically modified to increase its susceptibility to cancer, the cow modified to produce human insulin in its milk, and cloning to preserve or revive animal species. It is worth pointing out that individuals often reorganized their respective rankings following the group discussion.

Participants gave a variety of reasons why they found certain examples more appealing than others. Primarily these reasons included human health benefits (which, for many people, trumped food and other applications), greatest number of potential beneficiaries, and greatest benefit to or least instrumental use of animals. Other rationales for supportability included the least suffering for animals, maintenance of biodiversity, the least risk, environmental benefit, increased agricultural production, advancement of scientific understanding, and the greatest potential for public and regulatory acceptance.

Focus group participants consistently ranked two of the sample applications 'least supportable' more frequently than others. The genetically modified primate was completely unacceptable to most stakeholders, eliciting the strongest sense of revulsion that was articulated in different ways, mostly having to do with the proximity of the primate and the human (if the human brain tissue were grown in a lower animal, many were less strongly opposed). The other example that commonly drew a strong negative response was the glowing aquarium fish, which stakeholders perceived as 'frivolous,' 'trivial,' 'ridiculous,' 'gratuitous,' or as 'pure vanity.'

Following are some highlights from the individual focus groups. Participants in the *Scientists* group generally expressed high hopes for animal biotechnology, particularly for human health applications and for the overall advancement of scientific knowledge, but almost everyone drew the line at the genetically modified primate and the glowing aquarium fish. A chief concern was that the public might reject the technology due to unfounded fears, failed applications, and unbalanced, negative media coverage. Ethics was acknowledged as a major driver of public acceptability, and participants noted the need for scientists to make their voices heard in the ethics debate. Some members of the group raised concerns about considering values in public policy and regulation, in part because they thought that science shared a common truth, whereas values or ethics were heterogeneous and tend to fluctuate rapidly.

Participants in the *Animal Justice* stakeholder group were unanimous in rejecting animal biotechnology outright due to intrinsic objections to the instrumental use of animals. They argued forcefully that research using animals of any species is unconscionable, ineffective (not transferable to humans), and unjustifiable (as other alternatives could be pursued). Thus, they found none of the ten sample applications supportable.

Participants in the *Alternative Agriculture* group generally saw animal biotechnology as fundamentally flawed, but a few maintained a 'faint hope' that benefits might accrue to animals (relief from suffering, preservation of species) and to human health. The group's many moral and ethical concerns clustered around three main themes – political economy (biotechnology industries overwhelming developing-world, organic, and small-scale, sustainable agriculture), animal welfare, and environmental integrity.

Participants from both the Animal Justice and Alternative Agriculture focus groups raised concerns about the close relationship between regulators and industry and about the regulators' dual role as regulator and promoter of the technology. Overall, people in these groups tended to see government as a supporter of animal biotechnology, which lowered confidence in the regulatory system because human and environmental health was not always viewed as the government's primary concern. Several participants from these two groups wanted to see public values better incorporated into public policy and Canadian public opinion incorporated into the regulatory debate through a variety of deliberative mechanisms and through surveys, as well as improved public access to unbiased information on animal biotechnology. A more pluralistic range of stakeholders was desirable for policy discussions at the national and international level. Participants in these groups also recommended the incorporation of values further 'up-stream,' for example, by allowing granting agencies to assess the ethical aspects of proposed research projects before funding.

Participants in the *Agricultural Producers* stakeholder group were generally the most upbeat of all the groups about the potential for genetically modified food animals to increase agricultural productivity, create a new income stream for farmers, and provide protein for the developing world. However, participants expressed concern about regulatory hurdles, public acceptance, 'corporate control' of products and processes, and equitable distribution of benefits. One surprise was that several producers placed the example of the mouse genetically modified to have a susceptibility to cancer (which most participants in the groups found to be highly supportable) below their line of acceptability, owing to animal welfare concerns. Commented one farmer, 'I told you I was repulsed by some things and just the idea of a mouse that gets abused in that manner didn't turn my crank' (Agricultural Producers stakeholder V4).

The Agricultural Producers participants reflected a notably different stance on governance in comparison to those in the other six stakeholder groups. Overall, they agreed that regulation is critical for the commercial development of animal biotechnology products and for consumer confidence, but felt that Canadian society is over-regulated and that more decisions should be left to the market. Several participants insisted on the need for a governance framework that is void of public values primarily because societal values are seen to fluctuate and would cause too much uncertainty for the regulatory system. In addition, some producers argued that the export nature of Canadian agriculture meant that Canadian values were not so significant: regulators would need to consider global values instead. Participants generally wanted a science-based policy and regulatory system, with value decisions left to industry and the market, although they did not want to see a labelling regime in place. They argued that if consumers want to avoid animal biotechnology products, they could choose organic foods.

Participants in the *Health Care Providers and Patient Advocates* stakeholder group contributed some of the most nuanced discussion of ethical values and trade-offs. One participant spontaneously articulated a matrix of weighted criteria for assessing the acceptability of animal biotechnology applications. Most of the hopes for animal biotechnology were human health related, but always with the caveat that the application also benefit or do least harm to animals. Participants articulated little confidence in government's ability to provide leadership or oversight on such a 'huge undertaking' as animal biotechnology.

Health Care Providers and Patient Advocates focus group participants voiced frustration about the Canadian government's decision making. There was general agreement that government often commissioned reports but failed to act on the recommendations it received. Most participants wanted to see public values incorporated into policy decisions concerning animal biotechnology and wanted regulatory policy to have an 'ethical lens.' One person raised the example of citizen juries as a possible mechanism that would work better than surveys in garnering public values. Several people wanted to see industry, consumers, and the public at the table for government-led consultations on animal biotechnology policy and regulations. Participants from the Alternative Agriculture, Animal Justice, and Health Care Providers and Patient Advocates focus groups generally agreed that labelling was essential for the public to make informed decisions and wanted full disclosure so the public could know if the food product originated from a cloned or genetically modified animal.

Participants in the *Regulators* group noted that they are challenged to balance two needs: one, a supportive environment for scientific innovation, and the other, a regulatory review process that ensures public safety and confidence. People in this group were particularly concerned that their views not be generalized to make conclusions about a 'regulators' perspective' on animal biotechnology.

Participants in this group made it clear that regulators had to maintain a separation between science and values in risk assessment and that current regulations do not consider non-scientific concerns – science and safety are the ultimate basis for regulations. However, they felt that it was important to consider public values in the policy framework and that policymakers and regulators already effectively incorporate societal values in policy and regulations governing animal biotechnology, specifying societal values as human health and environmental safety. It was suggested that the democratic parliamentary system allows the Government of Canada to reflect societal values on a large scale and that public consultations manage to tease out societal values in the development of particular policies on animal biotechnology. Participants felt that industry needed to be responsible for providing consumer information on the food products derived from animal biotechnology and that industry should be building consumer acceptance – it was not the regulators' job to be providing labelling information.

Participants in the *Health Researchers* stakeholder group challenged one another in friendly debate in a well-balanced discussion of the benefits and risks of animal biotechnology. Overall, they supported the examples of human health applications, but drew the line at the genetically modified primate. However, participants made clear that genetically modifying animals was only supportable if the application had the potential to significantly reduce human suffering (e.g., to help find a cure for cancer or to create a safe, efficient, and economical supply of insulin) and could not be accomplished through any other means (as one researcher put it, 'We have no other way to answer the question' [V3]). Given these overriding criteria, the group had mixed reactions to food applications of animal biotechnology and a generally low regard for industrial and commercial applications.

Most Health Researchers focus group participants articulated a sense of confidence in the regulatory system, but suspected that enforcement was seriously lacking. In this group, concerns were raised about liability and equity, given the Supreme Court ruling in the *Schmeiser*[2] case, and about the lag between technology and regulations. Another view expressed was that regulatory policy needs to reflect public values, but that the public is uninformed and influenced by media spin that inaccurately or unfairly represented science. Representative and fair consultations that include the public were recommended as the best mechanism to incorporate public values. On the whole, participants suspected that the public would demand labelling of food products derived from animal biotechnology, but had concerns about the public's ability to understand labels, the usefulness of labelling a product 'GMO' or 'GM-free,' and the challenge of traceability.

It is important to remember that this summary of the focus group discussions cannot be interpreted as reflecting anything near a consensus of opinion within any one group on any of the issues presented to or raised by them. What the focus group discussions illustrate quite clearly, however, is that the particular 'stake' a person has in a technology like this one does have an influence on the values one brings to bear in thinking about the ethical and social policy issues the technology raises.

This influence happens in several ways. The obvious one is that one's 'stake' in a technology will have an impact on what one identifies as the most salient issues of that technology. Thus, the focus groups provided an important way of identifying aspects of the value questions around animal biotechnology that the authors of the chapters in this book may otherwise have missed.

The other impact of one's 'stake' in the technology is upon the way that values are prioritized. A fair conclusion to be drawn from the focus groups in this study is that the different perspectives and opinions on policy issues expressed by the participants were not, for the most part, generated by the fact that they held very different, even incompatible, values or beliefs. Instead, those different perspectives derived from a difference in the way the participants *ordered* these values. For example, the issue of the impact of this technology on animal welfare is a common concern – in fact, a surprisingly strong concern, even among those who might be expected to have the least concern, industrial producers – as illustrated by the concerns expressed by one farmer about designing mice for susceptibility to cancer. The producers, however, clearly did not prioritize this value in the same way the animal justice group did. In some cases, there was a surprising commonality in the prioritization of basic values – such as the widely shared rejection of the genetically modified glowing fish as a frivolously inappropriate use of the technology.

This suggests that the ethical debate around animal technology may not be as intractable as often assumed (as reflected in some of the focus groups). It does not appear to be a situation of incommensurate values, but rather one of value ordering. The latter is often influenced most strongly by disagreements about matters of fact, or by uncertainties about how technologies will develop and what their impacts will turn out to be. These are, in principle at least, more easily resolvable matters. What is required to resolve them are social institutions and practices that bring the disagreements into public debate and provide means of developing common understandings and consensus – or, at least, fair procedures of resolution. These questions are explored in the final chapter of this book.

NOTES

1 See the excellent discussion of the sense of 'abomination,' and its relation to the disruption of deeply embedded conceptual frameworks, in Stout 2001.
2 In *Monsanto Canada Inc. v. Schmeiser,* [2004] 1 S.C.R. 902, 2004 SCC 34, the Supreme Court of Canada ruled on the question of whether growing genetically modified plants constitutes 'use' of the patented genetically modified genome of the plant. The court ruled that the farmer who was sued by Monsanto had violated the patent rights of Monsanto by growing the crop in his field without paying the licensing fees to the company, whether or not he had planted the crop intentionally.

References

Fukuyama, F. 2002. *Our posthuman future: Consequences of the biotechnology revolution.* New York: Farrar, Straus and Giroux.

Gamborg, C., Gjerris, M., Gunning, J., Hartlev, M., Meyer, G., Sandøe, P., and Tveit, G. 2006. *Regulating farm animal cloning: Recommendations from the project Cloning in Public*. Copenhagen: Danish Centre for Bioethics and Risk Assessment.

Guardian Unlimited. 2008 (4 February). Shoppers care more about animals than climate. http://www.scenta.co.uk/nature/news/1715009/shoppers-care-more-about-animals-than-climate.htm.

Habermas, J. 2003. *The future of human nature*. Cambridge: Polity Press.

Holland, A., and Johnson, A., eds. 1998. *Animal biotechnology and ethics*. London: Chapman and Hall.

McKibben, B. 2003. *Enough: Staying human in an engineered age.* New York: Henry Holt and Co.

Pew Initiative on Food and Biotechnology [Pew Charitable Trusts]. 2006. Public sentiment about genetically modified food. http://www.pewtrusts.org/news_room_detail.aspx?id=32802.

Pollara Research and Earnscliffe Research and Communications. 2003. Public opinion research into biotechnology issues in the United States and Canada: Summary report prepared for the Biotechnology Assistant Deputy Minister Coordinating Committee. http://www.biostrategy.gc.ca/CMFiles/Wave_8_Summary_Report49RYS-922004 – 861.pdf.

Rollin, B. 1995. *The Frankenstein syndrome: Ethical and social issues in the genetic engineering of animals*. New York: Cambridge University Press.

Saad, L. [Gallup Organization]. 2004. The cultural landscape: What's morally acceptable? http://www.gallup.com/poll/12061/Cultural-Landscape-Whats-Morally-Acceptable.aspx.
Stout, Jeffrey. 2001. *Ethics after Babel: The languages of morals and their discontents*. Princeton: Princeton University Press.
Thompson, P. 1997. *Food biotechnology in ethical perspective*. London: Chapman and Hall.

2 Animal Biotechnology: The Scientific Landscape

MICKEY GJERRIS

The Science of Animal Biotechnology

What Is Animal Biotechnology?

One of the first things to realize about animal biotechnology is that it is not possible to give a simple answer to the question 'What is animal biotechnology?' The basic problem is that there is no agreement about where to draw the line in terms of what should be considered as biotechnology and what should be considered as more conventional uses of technology. Most people will readily agree that taking a gene from a human and inserting it into a goat embryo is an instance of animal biotechnology. But what about selective breeding? Or artificial insemination?

Behind these technical disagreements lies a value-laden discussion wherein, it is believed, the definition itself will help decide the ethical questions.

To many opponents of animal biotechnology, it seems that the more the definition stresses the novelty of the technology and the uncertainties about its effects, the more caution will be required to proceed. Not surprisingly, they therefore seek to have a narrow definition of animal biotechnology that includes only more recent developments, such as cloning and genetically modified animals. By contrast, the proponents of the technology believe that the more the technology is seen as a natural extension of already established practices, such as selective breeding and artificial insemination, the less reason there is to focus on the ethical aspects of the new technologies especially, since they are just a more advanced version of what we already know and accept (Clark and Whitelaw 2003). This discussion was also represented in the focus groups conducted within this project, where the arguments found in the literature were to a large extent repeated.

Now, combine this discussion about the nature of the technology with a discourse wherein proponents of the technology sometimes also stress the novelty of the technology and the limitless possibilities that it entails. Here the technology is often presented as something radically new. This most often happens when researchers or companies try to create attention and excitement about their research or products.

Clearly, the definition of animal biotechnology is not just a technical question, but is in itself a way of trying to promote a certain attitude towards the technology. Choosing what definition to work with is in itself a value-laden question. This illustrates well the very controversial nature of the technology. It is not even possible to pinpoint the subject without getting into the ethical discussion. And the question of definition also illustrates how facts and values cannot be separated from the beginning. Often there is a temptation to clarify the facts first and then discuss the values on the basis of these facts, as if the facts are neutral bits of information that can then be assessed through our value judgments. But facts are not neutral. This is not to say that something is not beyond discussion. Cells and DNA, cloning, and how to insert a gene into a genome are examples of factual realities that cannot be discussed. They are 'objective' in a pragmatic sense of the word. But how they are presented and how the discussion about them is framed is not a part of their objectivity. Already when we decide on whether to call something 'genetic manipulation,' 'genetic engineering,' 'genetic modification,' or 'genetic enhancement,' we apply our values to the world of facts.

In this book and in the focus group discussions, we have chosen to define animal biotechnology in terms of more recent developments within our understanding of molecular biology and genetics and our technological capabilities to utilize this knowledge. More precisely, throughout this book we will discuss animal biotechnology as technologies used to clone animals through some kind of somatic cell nucleus transfer (SCNT) or technologies used to modify the genome of an animal, either by inserting new genes into the genome or by 'knocking out' existing genes. We have done this for two reasons mainly. First of all, these are the uses of animal biotechnology actually discussed in the public ethical debate. While only a few discuss the negative impact of selective breeding on farm animals as part of the animal biotechnology discussion, the effects of cloning are a frequently discussed subject. Therefore, to capture the central elements in the public debate and the opinions of the various stakeholders, we have chosen to focus on the more controversial and novel examples of biotechnology.

Second, we have chosen this definition because it seems that although there is a continuation between, for instance, selective breeding in the traditional sense and cloning of elite animals for breeding purposes, there is also a categorical difference in the amount of power that humans take over the process of animal procreation in these cases. As the American animal ethicist Bernard Rollin has put it, we move from putting square pegs into square holes to changing the pegs and holes as we see fit (Rollin 2008).

So when discussing animal biotechnology in this book, we will mainly discuss cloning and genetic modification of animals through the means of modern biotechnology. We realize that this is not without problems, but we believe that if both we and the reader keep in mind that the whole area of animal biotechnology is ethically controversial, this necessary reduction of a very broad subject will not limit the ethical discussion to only some viewpoints, be they for or against the technology.

The Necessary Level of Knowledge

An important question when discussing the ethical issues related to animal biotechnology is how much knowledge one needs to have about the science behind the technology and the technology itself to be able to participate at an informed level in a debate on the ethical aspects of the technology. There is no simple answer to this question. There is no doubt that a certain amount of knowledge is needed, but there is also a risk of converting ethical considerations into technical subtleties if one overemphasizes the technical aspects. The American agricultural ethicist, and author in this volume, Paul Thompson discusses the issue and concludes that the level of knowledge needed is not as sophisticated as is often demanded in official and semi-official reports. He writes:

> While the capacity of the public for either unwarranted fear or enthusiasm should not be underestimated, it is quite questionable whether anything more than the most basic kind of science literacy is a prerequisite

> for beginning a discussion of ethical discussions in . . . biotechnology. One should know that scientists do not derive their theories by consulting oracles, of course, and one should have a vocabulary that makes sense of words like 'cell' and 'molecule.' Beyond this school science, one should know a few very basic things about genes and genetics. (Thompson 1997: 4)

It is furthermore important to remember that the level of knowledge needed to participate in a discussion about the ethical aspects of biotechnology is also dependent upon the specific topic under discussion. Basically, the discussion can be divided into three spheres: One focuses on risks: to human health, to animal welfare, and to the environment/ecosystem. A second focuses on the less tangible consequences to humans: the socio-economic impact of the technology and the psychological consequences that the ever-increasing domestication of nature have on humans. The last sphere focuses on the possible risks for harming the integrity, dignity, and naturalness of living beings that biotechnology might constitute. It should be noted that other subjects are discussed as well. The claim here is just that they will fit within one of the above-mentioned overarching spheres.

In the first sphere, the discussion is about the physical consequences of applying a certain kind of technology, and here there is the largest need for factual knowledge. If I am concerned that genes inserted into one animal (pigs) will spread into other wild-living animals (wild boars), these concerns can be clarified and the risk estimated to some extent by gaining knowledge about how genes are inserted into animals, how they are spread, what their function is, and so on. In the second sphere, the focus is on social and psychological consequences. Here the level of knowledge is not as decisive as in the first sphere. Basically, it is more important to understand what a transgenic pig will do to the relationship between small-scale farmers and more industrialized production systems within agriculture than to understand how to make one. In the last sphere, it becomes very complicated, because to state that something is unnatural can literally mean that it is found not to belong to nature. To make such a claim about certain animals or ways of producing animals, one needs to understand the technology and other ways that humans deal with animals. If it is unnatural to genetically modify a pig, is it not then also unnatural to use selective breeding? And if not, what is the relevant difference? On the other hand, the claim of unnaturalness can be a claim about a nagging feeling of control gone too far, of animals reduced to being meat factories on legs and an experience of something valuable being lost in the reduction of the animal to a technological challenge. Here it is not so much the technology itself that is debated as it is the attitude towards animals that it is an expression of. As one of the participants in the Alternative Agriculture focus group expressed it:

> We already sort of cram . . . poultry hens into cages in order to maximize the production. I mean, what are we capable of doing if we can get around some of those remaining biological constraints? . . . I think agribusiness reduces the value of animals to their profitability so I think we go as far as we could to maximize gain. (Alternative Agriculture stakeholder V3)

Finally, it is important to remember that animal biotechnology is an area with a wide variety of methodologies and applications. Thus, it can be hard to define the level of knowledge needed independently of the specific kinds of animal biotechnology.

To say it briefly: The context decides the needed level of knowledge. And in a book that seeks to encompass all the most relevant ethical considerations and concerns about animal biotechnology, that context is very broad. Therefore, the decision we have made is to give a brief introduction to the basic biological concepts involved in biotechnology and the most common methods used within animal biotechnology. We are fully aware that the level of knowledge needed to fully discuss and understand, let alone master, the technologies discussed in this book is not presented. Similarly, the level of knowledge needed to discuss specific risks related to specific kinds of animal biotechnology is not necessarily reached. What we hope to give is a very general introduction to the subject and the different applications that will enable the reader to follow the discussions about the ethical issues related to animal biotechnology.

This said, it is equally important to maintain that it is not necessary to understand all the technical details of animal biotechnology to be able to participate in a qualified way in the ethical discussion about the technology. Often it is more a question of being able to tell the difference between a technical disagreement and a value disagreement than of being able to discern the subtleties of the technical details.

Basic Biology. Molecular Biology: Cells, DNA, Genes, Proteins

As Charles Darwin (1809–82) already suggested, life has, in all likelihood, developed into its myriad forms from a single kind of organism dating back about four billion years (National Academy of Sciences 2002). This common ancestry of all life has great importance when understanding the possibilities within modern biotechnology. Many things separate the plurality of life forms we can see today at the level of everyday experience: they look, sound, feel, behave, and live in different ways. Some live in water, others in the air, some can move, others are immobile, some eat plants or meat, while others derive their energy directly from the earth and the sun. Some animals we feel we can relate to and have an idea of what is going on within their experience of the world, such as dogs, horses, and monkeys, while others, such as shrimp, fruit flies, and dung beetles, are much harder to relate to as other subjects. But if we look more closely at all these life forms, they begin to resemble each other more and more. And at the cellular and molecular level, it suddenly becomes obvious that all life forms are much closer related than they initially appear to be.

All living things are made up of cells. A human being is composed of somewhere between 10 trillion and 100 trillion cells. The rather large uncertainty is caused by both the fact that humans come in different shapes and sizes and the fact that cells are so small that it makes no sense to count all the cells in a human. So it is an estimate – and the most important lesson to learn here is that cells are very small and there is a fascinating number of them. Cells can have many different sizes, shapes, and functions. A human contains muscle cells, liver cells, blood cells, and so on. Although there are differences in the way that the animal and the plant cells are built and function, and even though there are huge differences in how humans and bacteria exchange DNA, there are also many similarities. Cells are what living things are made of. So although in the rest of this chapter we will limit ourselves to a discussion of animals, it is important to remember that in many respects cells are structured in the same way across the spectrum of living beings.

Two-thirds of a cell is made up of water, which means that two-thirds of the body is water. The rest of the cell is a mixture of molecules, mainly proteins, lipids, and carbohydrates. The cells turn the raw materials in the food we eat into molecules using thousands of different chemical reactions. The molecules are then used by the body to maintain its existence. But the cells are not just factories, because situated in the core of each cell is the genetic material that contains the 'recipes' for the molecules that the cells produce. This material is what we inherited from our parents. Half of our genetic material comes from the egg cell, the other half from the sperm cell: and that is all the genetic material needed to make a human.

The genetic material is organized within DNA molecules. DNA (deoxyribonucleic acid) molecules function as the storehouse of information for the organism. DNA is organized as the famous double helix that can look like a coiled zipper. The two strands of the zipper are composed of four molecules called bases. These are chemical compounds called adenine, thymine, guanine, and cytosine. These base pairs, identified through the first letter, A, T, G, or C, are the basic units in our genes. The order of the bases in a gene provides the recipe for building a specific molecule. The genes store the recipes in the DNA as a strand of the double helix. Other sequences of the same four bases tell the cell which gene to turn on, how much of the product molecule (protein) to produce, and where to produce it. The number of genes that code for a specific protein is debated, and estimates vary between 20,000 and 25,000 genes. The uncertainty in this area is not surprising if one considers that the human DNA (genome) is made up of approximately three billion base pairs and only a very small part of it is thoroughly understood. It is estimated that only about 5 to 10 per cent of the DNA is actually part of genes that code for something. What the rest is doing is largely unknown. This has not, however, prevented it from being labelled 'Junk-DNA.' As our knowledge of the way DNA is organized and how it unfolds in the organism increases, we are gaining a better picture of the role of at least some of this seemingly irrelevant information lying dormant in the cells.

If we leave aside all the material in the DNA that is not transcribed by the cell into recipes for making proteins (the Junk-DNA), we are left with the basic information carrier in living beings – the gene. It is now that the common ancestry of all living beings becomes very interesting, because genes have the same structure in almost all living organisms. Humans, mice, trees, lichen, and bacteria store information in much the same way. Because genes are similarly organized, they are readily moved from one organism to another. This means that human cells are capable of transcribing a gene whether it comes from a human or a mouse. It looks the same to the cell. And it is therefore possible, theoretically, to take any gene from any organism and put it into the genome of another organism and have the cell produce the protein that the gene contains the recipe for.

Our knowledge of how the genes interact with each other and the environment is growing fast. Growing just as fast is our understanding of how little we know. Or to put it more bluntly: Things are much more complicated than was originally envisioned. We will just mention two areas where the fundamental understanding of what genes are and how they work is changing.

First, it is important to be aware of the importance of *epigenetics*. Broadly understood, the term describes all the factors besides the DNA sequence present in cells that influence the development of the organism. This includes both influences from higher layers of organization in the organism itself and influences from the environment. Understood more narrowly, epigenetics refers to heritable changes in gene expression caused by mechanisms other than changes in the underlying DNA sequence. This means that the DNA sequence stays the same, but the organism changes nonetheless. These changes may remain through cell divisions and may also last for multiple generations (Bird 2007).

Second, the whole idea of genes as separate carriers of information is changing since it has become more and more clear that everything is interconnected – also at the molecular level. Changes in the DNA sequence in one gene can very well lead to changes in the function of other genes. The boundaries between genes seem to become less and less distinct the more we learn, and our knowledge of the complexity of the genomic interactions is growing. What we are left with today is a much more abstract, open, and general concept of genes than previously held (Portin 2000).

If we leave aside this emerging understanding of the seemingly ever-growing complexity of living things, we can sum things up as follows: All life forms are built out of cells. And in the core of the cell we find the genetic material that contains the recipes for the compounds that our bodies need to work. The genetic material is organized in DNA molecules that are made up of small chemical building blocks. As far as we can see, only some of this material is used for storing information that is used by the body. If we have the technical abilities, we can move the genetic material between all species, thus enabling one organism to produce the compounds encoded in another kind of organism's genome. It is the practical aspects of this that animal biotechnology is all about. But before going into more details about the applications of animal biotechnology, there are a few more technical aspects that need to be clarified.

Genetic Modification and Cloning: What Are They?

An animal that has had its genome changed is usually called a 'genetically modified animal.' The changes in the genome can be of several kinds. The most spectacular is undoubtedly the one whereby a gene from one kind of organism is placed in the genome of another kind of organism. Such an animal is called a 'transgenic animal.' Less dramatic, but no less influential, changes in the genome can be reached by turning off or removing existing genes in the genome (knockout animals), inserting extra copies of genes already present, or inserting genes from closely related animals (cisgenic animals). Finally, it should be mentioned that the genes inserted into the genome can also be artificially synthesized.

There are several ways to introduce these changes into the animal genome. The most common way is called *pronuclear microinjection*. Here the relevant DNA construct is injected into the core of a fertilized egg. The method is rather easy to use, but the efficiency is low, since one can only hope that the string of DNA will insert itself into a place in the existing genome where it does not disrupt other genes and where it can work (be expressed). This method has been used to create transgenic mice, sheep, cattle, pigs, goats, and rabbits, albeit only 3 to 5 per cent of the animals born carry the desired transgene (Hunter et al. 2005). Another way is *sperm-mediated DNA transfer*, whereby the genetic changes are introduced in the sperm cell before fertilization of the egg. There are several ways of getting the genetic material into the sperm cell, but again they suffer from inefficiency (Robl et al. 2007). The most promising method of introducing genetic changes is the *viral vector transfer*, whereby the DNA construct is transported into the cell genome by a genetically modified retrovirus. Retroviruses are known for their ability to insert themselves into the genome of the host, and it is this ability that researchers seek to utilize. They seek to modify the retrovirus so that the actual virus part (what makes you ill) is no longer present and then utilize the carrying capacity of the retrovirus to carry a different load: the DNA construct that is to be inserted into the genome of the host cell. This is done in cultured cells that are then inserted either into a non-fertilized oocyte that is afterwards fertilized or into a zygote (an embryo at a very early stage of its development) before it is transferred to the surrogate mother (Hunter et al. 2005).

These are all ways to produce animals with genomes altered by having something inserted into them: a transgene (a gene from a wholly different species), a cisgene (a gene from a closely related species), an extra copy of a gene from the same species, or an artificially constructed gene. Knockout animals, by contrast, are animals where one or more genes have been disabled. This can be done either by using a pair of molecular-sized 'scissors' to cut out a well-defined part of the genome or by inserting a gene into the original gene, thus disabling its ability to express itself. Knockout animals provide valuable information on the role of a specific gene by enabling researchers to study what happens if the gene is not functional. To produce a knockout animal, one must first develop genetically altered cells, which are then transferred to embryos that, after birth and further breeding, create knockout animals. This technology was developed in the 1980s for mice (Clark and Whitelaw 2003), but only in the last few years has the technology been developed for other animals, most notably pigs (Milland et al. 2005), but also sheep (Denning et al. 2001).

The final way to produce genetically modified animals that should be mentioned here is through cloning technology. Since the presentation of the Dorset ewe Dolly back in 1997 by Ian Wilmut and other researchers at Roslin Institute in Edinburgh (Wilmut et al. 1997), this is unarguably the animal biotechnology that has led to most controversies and captured most minds. And, since cloning can be a helpful tool in producing genetically modified animals, we will now turn to this technology.

Clones are almost exact genetic copies of each other. There are many examples in nature of cloning understood as organisms that are either genetically similar (like identical twins) or who procreate by cloning (potatoes). In this connection, though, and in line with our focus on newly developed animal biotechnologies, we will focus on the method known as *somatic cell nuclear transfer*. Here the nuclear DNA from a cell isolated from an adult individual is inserted into an oocyte that has had its original nucleus removed. The oocyte is then manipulated into behaving as if it had been fertilized by a sperm cell, thus beginning to divide. The point is that the genome in the new organism is almost the same genome as that of the original adult animal.

When we write *almost,* it is because there are some subtleties that are necessary to explain. When we said earlier that the genetic material of the cell was located in its core, this was a truth needing a small modification. For in the outskirts of the cell where the proteins are produced on the basis of the recipes encoded in the genes, there is a little bit more genetic information. This is known as mitochondrial DNA. The role and function of this material is still being explored, but it seems to play an important role in the development of the organism. This genetic material is not 'copied' in the cloning process, and so the clone has an almost identical genome as the original animal but for the mitochondrial DNA, which in humans span about 16,500 DNA base pairs containing 37 genes, a fraction of the total DNA in cells (Vajta and Gjerris 2006). However, it is not quite clear how much of the mitochondrial DNA from the oocyte cell and how much from the donor cell is utilized during the development of the fetus (Ferreira et al. 2007). This is one of the areas within cloning where further research is needed to gain an understanding of the biological processes involved.

The difference in the mitochondrial DNA and the epigenetic factors mentioned earlier that influence the individual from the point of fertilization are seen as some of the main reasons why the cloning technology still struggles with relatively low efficiency. Different species such as mice, rats, rabbits, sheep, goats, pigs, cattle, horses, cats, and dogs have been cloned over the past ten years, with success rates remaining somewhere between 0.1 and 20 per cent, depending on the species. Among the epigenetic factors are the kind of nutrition fluid the fertilized egg grows in, the diet of the surrogate mother, and the compounds the fetus and the born animals are subjected to. These differences in the environment (broadly understood) between the original animal and the cloned animal are seen as an important part of the explanation of the phenotypic difference (the difference on the organism level) between clones that are almost identical seen from a genotypic level (the genetic level).

Cloning can be used as a technology in itself to create almost identical copies of existing animals. But it can also be used in conjunction with genetic modification by providing a tool to produce transgenic animals by cloning them or by introducing the desired genetic changes into cells, cloning them, and then when the oocyte has begun to divide into an early embryo, using these cells for further modification, thus enabling a whole series of changes to be made to the cells. This work is only just beginning, and it is still uncertain whether it can be successful or if epigenetic effects will make it impossible to control the changes in the cells (Robl et al. 2007).

To conclude this section, it must be emphasized that this overview of the scientific and technical aspects of animal biotechnology is in no way meant to cover the whole area. And as animal biotechnology is a rapidly developing field within science where the understanding of the genes and the way they interact with the environment grows all the time and where methodologies for introducing genetic material into cells or cloning them are refined or replaced with astonishing speed, one cannot expect the knowledge presented here to be up to date on the technical details for long. Rather, the goal here has only been to provide the reader with enough knowledge to be able to follow the discussions about the ethical aspects of animal biotechnology in this book.

The Applications of Animal Biotechnology

The Context of Animal Biotechnology

Animal biotechnology is not an isolated human activity. The motivation for developing the technology is the same as that behind other technologies, both new and old, and the social context, the social reality that the technology develops in, is the context and reality of our everyday lives. It is, in all meaning of the words, a very human activity. Not surprisingly, it exhibits many of the traits that our other activities do. It is ambiguous, can be misused, and has great potential. There is no room in this book to go into details about the social context of the technology or how it relates to the world of science in general. Instead, we have chosen to highlight a few points that are important in the discussions that follow.

Humans are vulnerable and mortal animals, a fact we tend to overlook. This is probably because it is very hard to live with the constant knowledge that we have to part one day from everyone and everything we love. Nonetheless, our lives are lives 'in cura' as the German philosopher Martin Heidegger (1889–1976) put it. Our lives are lives 'in worry,' to translate the phrase freely. We are constantly challenged by our very existence to maintain the same existence. We have to find food, water, shelter, clothes, human companionship, and rest to survive. We have to maintain our lives. This might be possible to forget in the more wealthy parts of the world today, where the problem often is that people eat too much. But for many human beings, life is still a daily struggle to provide the necessary elements for the continuation of a human life.

If we look for the reasons behind, not just animal biotechnology, but technology in general and perhaps even wider to the reasons behind the continued human struggle to understand and control our surroundings, this is where we end – in a human existence that is from the outset laid out in such a way that understanding and controlling the world is a way of ensuring life. Animal biotechnology thus becomes but one of the latest chapters in the long story of how humans have tried to become masters of a world that at the same time provides all that is necessary for life but also, through sudden changes in climate and populations, for example, threatens to take it away.

Animal biotechnology is one form of this control – control over animals in order to make them better providers of the benefits we need from them, be it knowledge, medicine, food, or luxury. But at the same time, animal biotechnology is a technology developing within a certain type of society. In our case this is a capitalistic society where there is a long history of treating animals as commodities. And even though the past thirty to forty years have seen changes in the general outlook on animals and improvements (from the animal's perspective) in welfare, it is clear that animal biotechnology is developing in a culture where it is already considered legitimate to treat animals as things and means to our ends.

Animal biotechnology is also a business where the legitimate goal is to earn money. Thus, the motivation behind developing a certain kind of animal biotechnology might be hard to discern. It can be the expression both of a will to control nature in a way that will help ensure human life, but also of the relentless desire to make profits, even when there is no actual need for the products developed. Most often it is a mix of these. One of the participants in the Alternative Agriculture focus group expresses it in this way:

> I don't have a concern per se with the science itself, I really believe that . . . humans should push these boundaries of scientific exploration but I have more of a problem with the . . . commercializing [of] the results, and I think that's where things begin to go awry because that is where . . . something very natural called human greed comes into play and especially manifested through capitalism and our system of . . . our economic system that I think has given a lot of power to corporations which . . . exist to maximize short-term profit. So you are turning . . . an invention or an advancement in scientific thinking – it's in the commercialization

process where I think the problems happen. (Alternative Agriculture stakeholder V3)

It is worth remembering that animal biotechnology is developed by humans motivated by the same factors that motivate most of us: the desire to help those less fortunate than ourselves, the desire to live good lives, and the desire to earn social respect and prestige. In other words, researchers are humans who have a job. And they even have highly competitive jobs. The world of research today is a highly competitive environment where there is a constant struggle to provide results, publish articles, and attract funding. It is therefore hard to find researchers who underestimate the possible impact of their research. On the contrary, expectations are usually rather high. Sometimes they are so high that it might be more accurate to talk about *hype* or *over-selling*. This simply means that the promises made on behalf of the technology do not always hold true or, to phrase it more sympathetically, might be overly optimistic.

All this should be kept at the back of one's mind while reading both the remainder of this chapter and the rest of the book. Animal biotechnology is embedded in a social context and therefore related to everything human. The motives for doing animal biotechnology can be very hard to discern in the individual cases, but one should neither be overly optimistic about the altruistic motives behind it or overly suspicious about possible egoistic and narrow motives. It is often a very human mixture of these elements. And, when hearing of all the problems that the technology might solve and how soon it will revolutionize our lives, one should always take into account that the worlds of research and business, of science and technology, of knowledge building and knowledge using, have melded together in our Western capitalistic society. So scientists are no longer necessarily independent and altruistic, but are humans working in a marketplace like everybody else. They are, in other words, stakeholders. A certain amount of sound scepticism in evaluating the claims they make is therefore probably not wasted.

That said, it is time to look at what animal biotechnology can do and what it is hoped it will be able to do in the future. The potential is enormous and the range of applications almost endless – in theory. But as the last twenty-five to thirty years have shown, there is a big leap from having an idea to making it work. Sometimes it seems that the more we learn through science, the less we know, because we keep finding out that the world in general and molecular biology specifically is much more complicated than we initially believed. As mentioned above, it is not an easy job to evaluate the claims about what will be possible and when a certain application will be available, but on the basis of the current literature, we have tried to conduct such an evaluation.

The Possible Applications

As we discussed earlier, all living beings share the same ancestry and therefore share the same way (or at least very similar ways) of storing and utilizing information in the cell. This means that genetic information from one organism can be moved to another organism, thus opening up new opportunities. At the same time, it is now possible to copy almost an entire genome from one organism into a new organism, thus producing a clone that will share almost all genetic characteristics with the original. The possibility of modifying an animal genome in one way or the other, taken together with the possibility of cloning, provides animal biotechnology with a wide range of opportunities. What we have done in the following is to divide the applications into different areas and in each area give some key examples of what the technology can be used for.

RESEARCH

Understanding biological mechanisms, such as how the cell transcribes the information in the genes and turn it into proteins, how the genome is copied each time the cell divides itself, what it is that regulates what genes are turned on and off, what the relation is between the amount of oxygen a fertilized egg 'inhales' and the changes that it will end up with as a viable individual, is an inescapable challenge if biotechnology is to fulfil the hopes and expectations it has raised. The deeper the understanding of the mechanisms that one tries to utilize through technology, the higher the chances of success. Basic biological research is therefore a necessity in order to discover the possibilities and limitations that a technology entails. As an example, the scientific understanding of cloning has changed in the past ten years as more and more research has shown how copying a genome into a cell does not lead to an exact copy of the 'founder' animal. Both the role of the mitochondrial DNA and that of all the epigenetic effects have been shown to be more significant than initially expected (Vajta and Gjerris 2006).

Understanding how the genes are distributed in the DNA and figuring out what role the different genes play in the development and existence of the organism is another area in which basic research plays a very important role in realizing the potentials of animal biotechnology. Laying out gene-maps that show how the bases are distributed along the DNA strand (called sequencing) and learning how to interpret what combinations of base pairs to look for to find sequences of the genome that contains a gene that encodes for a specific protein are thus central elements in animal biotechnology.

Research can be both targeted at a specific problem and driven by a very detailed hypothesis that is to be tested. It can be broad and basically just try to describe an area, then look for patterns or rules that can be used in understanding the phenomena at hand. Biotechnology research is often relatively specific since, to a large extent, the funding comes from private investors who have an interest in utilizing the technology within a relative short time span. But it is important to realize that the specific or applied research cannot succeed without basic knowledge. Thus, it is important that both kinds of research are funded if science is to progress. Without basic understanding and the possibility of asking fundamental questions, the sources of knowledge that open up the more specific potentials dry up. Therefore, the development of animal biotechnology is also embedded in the complex relationship between public and private research, research funding, and the discussion of the role of university research in an increasingly business-inspired society.

MEDICINE

The area where animal biotechnology has so far played an important role, and in all likelihood will continue do so in the future, is medicine. Here, both genetically modified and cloned animals can be used in a number of ways.

First of all, cloning can be used to produce genetically very similar animals for experimentation and drug testing. The similarity of the animals could exclude some sources of uncertainty as researchers would know the genetic make-up of the research animals. This could be useful both in animal experiments in general and especially when testing new drugs for toxicity. And it could, theoretically, reduce the number of animals needed for experimentation.

The cloning technology becomes even more useful when used as an assisting technology in connection with the production of transgenic animals. Here two possibilities are most prevalent: animals as models for human diseases and animals as producers of human proteins. In the first case, the animals are genetically modified so that they exhibit the characteristics of a human disease. A famous example in this area is the *oncomouse*. Produced by Harvard University and DuPont in the mid-1980s, this mouse was genetically modified to carry an activated cancer gene that significantly increased the mouse's susceptibility to cancer (Stewart et al. 1984). The oncomouse was used as an example of animal biotechnology in the focus groups and was the sample application that drew the greatest overall support from the participants. To date, literally tens of thousands of animal models, mainly mice, have been developed. Diseases such as cancer, Parkinson's, Alzheimer's, diabetes, and cystic fibrosis are all studied using animal models. One of the latest attempts to utilize the technology is the so-called Alzheimer-pigs research produced in Denmark. Here, pigs have had genes inserted that are believed to be responsible for Alzheimer's disease in humans. The hope is that the pigs will exhibit traits similar to those of Alzheimer patients and thus both enable a deeper understanding of the disease and provide the researchers with animals that are well adapted for the testing of new drugs (University of Copenhagen 2007).

The other application that should be mentioned here involves *pharm* animals, which are genetically modified to produce human proteins, typically in their mammary gland, and monoclonal antibodies, typically in the blood, thus enabling the retrieval of the protein or antibody from the relevant tissue. This is seen as an alternative to the use of chemical synthesis or microbial and mammalian cell culture systems. Today, one drug is on the market: antithrombin III (ATIII), an anticoagulant produced from the milk of transgenic goats and used for the treatment of certain patients in connection with heart operations. Many other drugs are currently being developed this way, and expectations are that the market for such proteins will exceed several billion USD in a few years (Niemann et al. 2005). The technical difficulties in producing transgenic animals that express the right genes in the right way should not be underestimated. The initial costs in developing an animal that can function as a *bioreactor* are high, but if and when the system works and the drug is approved, it could prove to be a low-cost alternative to the more conventional production systems.

One application of animal biotechnology that is often mentioned as a reason why the technology should be allowed to progress despite ethical controversies is *xenotransplantation*. The vision is to be able to produce animals, typically pigs, that can be used as organ donors, thereby solving the problem of a shortage of human donors. The idea to use animals as donors of organs has been around since the 1960s, and the technological development is still far from complete.

To succeed, there are three problems that need to be solved: first of all, the serious immunological problem that the human body will reject an organ from an animal. Researchers have succeeded in producing pigs that have been genetically altered in a way that, in theory, should 'fool' the human body into believing it is an organ from a human (Yamada et al. 2005). Whether it will work in 'real life' is still too early to say.

A second reason that it will take some time before such experiments can be successful is that the pig genome contains genetic diseases called *porcine endogenous retroviruses* that could be transferred to humans if pig organs are introduced into the human body. The risk that this transfer will happen is probably not very high, although researchers disagree as to how high the risk is. But because a disease that lies dormant in the pig genome could come to life in a human body and develop into an epidemic like AIDS or SARS, the risk does not need to be very high to be deemed extremely serious (Martin et al. 2006).

The third requirement for the use of xenotransplantation in human treatment would be to design the pigs in a way that is compatible with human anatomy and physiology.

Xenotransplantation is a good example of how animal biotechnology is embedded in a cultural discussion about ethical and social issues. This technology is almost always mentioned by proponents of animal biotechnology as one of the reasons why we should move ahead in developing biotechnology for medical uses. The technology is described as close to being ready and as entailing only manageable risks (Niemann et al. 2005), whereas more critical commentators on animal biotechnology see it as more of a visionary technology that, if problems 1 and 3 are eventually solved, would still entail risks unmanageable and serious enough for it not to be used (Gjerris and Sandøe 2009). To give a realistic estimate of the future of this application of animal biotechnology is very hard. And both sides have compelling arguments.

Its proponents can promise to solve the organ shortage problem and to prolong human lives, whereas its opponents can point to scientific uncertainty and the possibility of disastrous pandemics. Who is right can only be known in the future. In the present, however, the technology, still in its infancy, nonetheless plays an important role in the discussions about animal biotechnology in general.

Whether xenotransplantation of organs from pigs to humans will ever come about or not, there is no doubt that a lot of applications of animal biotechnology will be developed in the area of medicine. In particular, the control of nature that biotechnology provides and the similarity between humans and other animals at the genetic level taken together open up a lot of opportunities for understanding and treating diseases in new ways. These opportunities, combined with the will to spend large sums of money on medicine that can prolong human life, make the area of animal biotechnology of interest from both a humanistic and a commercial perspective.

AGRICULTURE

Theoretically, any gene from any organism can be expressed in any organism. The range of applications of animal biotechnology is therefore almost endless. In agriculture, the following opportunities are usually mentioned in the literature: increased production, decrease in costs, changes in product composition, reduced environmental pressure, and increased animal welfare.

The cloning of valuable breeding animals is one way that technology might increase production. It could, for example, enable a farmer to clone an elite animal and gain double the offspring from it. Whether this is economically viable with the current success rates in cloning is questionable, but in the longer term this method is expected to be useful in some productions systems, for instance, aqua-farming and cattle breeding. However, it is worth mentioning that the limitations of cloning, specifically the fact that only almost exact copies can be produced, and that cloning actually blocks the progress that would be gained through traditional selective breeding, could reduce the technology's usefulness (Gjerris 2006).

Another way to utilize the technology would be to decrease the costs of production by improving the food conversion rate of the animals, changing the body/fat ratio, or even changing the nature of the product by, for example, making pigs with increased amounts of non-saturated fatty acids or milk from cows that contains the necessary amounts of human lactoferrin to make it instantly usable for babies. Such animals have been produced in the laboratory, and it seems only a matter of time before they will enter the marketplace (Niemann 2005). Again, it is important to see that although it might be very expensive to develop an animal that basically grows faster with more muscle mass on less food, or an animal that produces a product with more desirable traits than a conventional animal does, the potential profits are also very high as the whole goal of animal production is to get as much out of the animals, with as small an input, as possible. Growth rates and food conversion efficiency have improved dramatically since the advent of organized breeding systems in the 1950s. It is hoped that animal biotechnology can take this development further. Of course, there is also a downside, as the increasing pressure on the animals has caused severe welfare problems in many cases, but we will return to this in the next chapter as well as looking at the social scepticism towards these applications of animal biotechnology that to some are the main reason that the technology is not already on the market shelf (ibid.).

An example of an attempt to use the technology to produce an animal that puts less pressure on the environment is the Enviropig™ developed by researchers at the University of Guelph (Forsberg et al. 2003). This pig has been genetically modified to carry a gene from a bacterium in its salivary gland enabling it to digest plant phytate. In conventional pigs the phytate phosphorous goes undigested through the pig and back into the environment, where it pollutes and puts pressure on water reserves. Pigs that have been genetically modified to carry the enzyme have been shown to reduce the amount of phosphorous in their manure by up to 75 per cent. Furthermore, the conventional pigs need to have phosphorous added to their diet, which is an additional cost to the farmer. Thus, the Enviropig™ seems to reduce both the environmental pressure of pig production and the costs to the farmer. It is one of the agricultural applications which seems to be closest to the market, although there is no saying exactly when the animal will pass through the necessary regulatory approval procedures, let alone whether it will be seen as an ethically acceptable use of the technology and gain consumer approval.

The last application that should be mentioned is the possibility that animal biotechnology might be used to solve some of the welfare problems imposed on animals over the last fifty years as increasing pressure has been placed on breeding for higher production. Many kinds of farm animals, including slaughter chickens and dairy cows, have severe welfare problems in the conventional breeding systems. Chickens suffer from leg problems due to growth rates that are too swift for their bone structure, whereas dairy cattle suffer from mastitis due to the high production of milk that they have been bred for. This is not only painful to many of the animals, but also a factor that limits production. Such problems could potentially be solved by genetically modifying the animals to 'fit' better into both their growth rates and the production systems. Finally, it should be mentioned that many people envision that animals could be genetically modified to be resistant to certain infectious diseases that are not only painful and often lethal, but also threatening to the farmer, such as swine fever and Newcastle disease in poultry (Clark and Whitelaw 2003). This could, as one of the participants in the Regulators focus group noted, benefit both humans and animals: 'If you had enhanced resistance to disease of animals which can affect people, like cattle (that) are more resistant to mad cow disease, that would [be] safer as a source of food. If you could enhance their overall health by enhancing resistance to specific diseases, then that would be transmitted as a benefit to humans' (Regulator V6).

The applications of animal biotechnology within the agricultural sector are still only at the experimental stage. As with the other applications of animal biotechnology, it is hard to say exactly when the animals will leave the labs and enter the marketplace. But it seems safe to say that the agricultural applications are more controversial than the medical ones. We will return to this point in the next chapter. However, it is worth mentioning that the success of animal biotechnology in this area is dependent not only upon the technology being useful and economically viable, but also on whether it is seen as acceptable by society at large. Otherwise, it may make little sense to put the products 'on the shelves.'

Conclusions

Horses that are popular within certain kinds of sports have been cloned, just as companies are offering to clone cats and dogs. Fish have been made that glow in the dark, initially to help in controlling pollution in rivers, but eventually as aquarium novelties. An artist has made a real, glowing bunny, the Chinese hope to use the cloning technology to save the panda, Russian researchers work on reviving the mammoth, and attempts are made to produce a genetically modified non-allergenic cat. Each day seems to bring a new story. What can be said here is that although all the above-mentioned examples are taken from the real world, a lot of hype and easy headlines are seen in this area. Biotechnology is not like building with Lego. It is a technology in its early stages that is expensive, has low success rates, entails experimentation on living beings that can experience pain and welfare problems, and is ethically controversial. It is a technology that is making it more and more clear just how complex and interwoven biological creatures are. All this means that although individual scientists might be tempted to get a minute of fame in the limelight, and although a reporter might be tempted to get an easy story, it is a long way from the promises of the sensational press to the reality of the laboratories where the science and technology is being done.

There is no doubt that even more colourful examples of animal biotechnology than the ones mentioned above will see the light of day and make the front page in the years to come. But it seems equally certain that this will not be in the areas where the technology will play the largest role. Rather, it will be within research and medicine that animal biotechnology will have the greatest influence. What will happen within the area of agriculture is hard to say. That depends not only on the technologies becoming more efficient and economically viable, but also on their social acceptability. This issue will be discussed elsewhere in this book (see the chapters by Ries and Sheremeta). For now it suffices to say that so far, it seems, it will be hard to convince ordinary citizens that we need genetically modified animals on our dinner plates. But things can change and have a tendency of fooling those who make prophecies.

References

Bird, A. 2007. Perceptions of epigenetics. *Nature* 447: 396–8.

Clamp, M., Fry, B., Kamal, M., Xie, X., Cuff, J., Lin, M.F., Kellis, M., Lindblad-Toh, K., and Lander, E.S. 2007. Distinguishing protein-coding and non-coding genes in the human genome. *Proc. Natl. Acad. Sci. USA*, 10.1073/pnas.0709013104.

Clark, J., and Whitelaw, B. 2003. A future for transgenic livestock. *Nature Review Genetics* 4: 825–33.

Denning, C., Burl, S., Ainslie, A., Bracken, J., Dinnyes, A., Fletcher, J., King, T., Richie, M., Ritchie, W.A., Rollo, M., de Sousa, P., Travers, A., Wilmut, I., and Clark, A.J. 2001. Deletion of the Alpha (1, 3) galactosyl transferase (GGTA1) gene and the prion protein (Prp) gene in sheep. *Nat. Biotechnol.* 19: 559–62.

Ferreira, C.R., Meirelles, F.V., Yamazaki, W., Chiaratti, M.R., Méo, S.C., Perecin, F., Smith, L.C., and Garcia, J.M. 2007. The kinetics of donor cell mtDNA in embryonic and somatic donor cell-derived bovine embryos. *Cloning and Stem Cells* 9(4): 618–29.

Forsberg, C.W., Philips, J.P., Golovan, S.P., Fan, M.Z., Meidinger, R.G., Ajakaiye, A., Hilborn, D., and Hacker, R.R. 2003. The Enviropig physiology, performance, and contribution to nutrient management advances in a regulated environment: The leading edge of change in the pork industry. *Journal of Animal Science* 81: E68–77.

Gjerris, M. 2006. *Ethics and farm animal cloning: An examination of the risks, values and conflicts related to farm animal cloning*. Denmark: Danish Centre for Bioethics and Risk Assessment.

Gjerris, M., and Sandøe, P., 2009. Transgenic animals. In *Encyclopedia of environmental ethics and philosophy*, ed. J.B. Callicott and R. Frodeman. Detroit: Macmillan Reference USA.

Hunter, C.V., Tiley, L.S., and Sang, H.M. 2005. Developments in transgenic technology: Applications for medicine. *Trends in Molecular Medicine* 11(6): 293–8.

Martin, S.I., Wilkinson, R., and Fishman, J.A. 2006. Genomic presence of recombinant porcine endogenous retrovirus in transmitting miniature swine. *Virology Journal* 3: 91.

Milland, J., Christiansen, D., and Sandrin, M. 2005. [alpha] 1, 3-Galactosyltransferase knockout pigs are available for xenotransplantation: Are glycosyltransferases still relevant? *Immunology and Cell Biology* 83(6): 687–93.

National Academy of Sciences (NAS). 2002. *Animal biotechnology: Science-based concerns*. National Academy of Sciences, Washington, DC.

Niemann, H., Kues, W., and Carnwath, J.W. 2005. Transgenic farm animals: Present and future. *Rev. Sci. Tech. Off. Int. Epiz.* 24(1): 285–98.

Portin, P. 2000. The origin, development and present status of the concept of the gene: A short historical account of the discoveries. *Current Genomics* 1(1): 29–40.

Robl, J.M., Wang, Z., and Kasinathan Pand Kuroiwa, Y. 2007. Transgenic animal production and animal biotechnology. *Theriogenology* 67: 127–33.

Rollin, B.E. 1996. Bad ethics, good ethics and the genetic engineering of animals in agriculture. *J. Anim. Sci.* 74: 535–41.

– 2008. Commentary, Animal ethics and the law. *Michigan Law Review First Impressions* 143.

Stewart, T., Pattengale, P., and Leder, P. 1984. Spontaneous mammaryadeno-carcinomas in transgenic mice that carry and express MTV/myc fusion genes. *Cell* 38(3): 627–37.
Thompson, P.B. 1997. *Food biotechnology in ethical perspective*. London: Blackie Academic and Professional.
University of Copenhagen. 2007. Cloned pigs help scientists towards a breakthrough in Alzheimer's. *Science Daily*. http://www.sciencedaily.com/releases/2007/06/070629083416.htm.
Vajta, G., and Gjerris, M. 2006. Science and technology of farm animal cloning: State of the art. *Animal Reproduction Science* 92: 210–30.
Wilmut, I., Schnieke, A.E., McWhir, J., Kind, A.J., and Campbell, K.H., 1997. Viable offspring derived from fetal and adult mammalian cells. *Nature* 385(6619): 810–13.
Yamada, K., Yazawa, K., Shimizu, A., Iwanaga T., Hisashi, Y., Nuhn, M., O'Malley, P., Nobori, S., Vagefi, P.A., Patience, C., Fishman, J., Cooper, D.K., Hawley, R.J., Greenstein, J., Schuurman, H.J., Awwad, M., Sykes, M., and Sachs, D.H. 2005. Marked prolongation of porcine renal xenograft survival in baboons through the use of alpha1, 3-galactosyltransferase gene-knockout donors and the cotransplantation of vascularized thymic tissue. *Nat. Med.* 11: 32–4.

3 Animal Biotechnology: The Ethical Landscape

MICKEY GJERRIS

Why Are We Discussing Ethics All the Time?

In 1997, the Dorset ewe Dolly was presented to the world by a group of researchers led by Dr Ian Wilmut at the Roslin Institute in Edinburgh. Normally sheep do not give rise to headlines in media around the world, but Dolly did. She was a clone, a close genetic copy of an adult animal. She was produced by taking a cell from the mammary gland of an adult sheep and fusing it with an egg cell from another sheep that had been emptied of the genetic material in the cell core. This produced a fertilized egg that was transferred to a surrogate mother, and after a normal pregnancy, Dolly was born. This was something widely believed until then to be biologically impossible in mammals.

So Dolly hit the headlines. Not only because of the scientific excitement, but also because of the new technological possibilities that opened up. If sheep could be cloned, could not humans? The general agreement was that it was only a matter of time before the technical prerequisites to clone humans were available, but that this would be ethically wrong. What was only discussed rather fleetingly in the media at that time was, however, what has so far turned out to be the most important use of the technology: the cloning of animals for a wide spectrum of purposes. Cloned animals are now, more than ten years after the birth of Dolly, beginning to emerge from the labs and into the public sphere in different areas. And whereas there is almost unanimous agreement that reproductive cloning of humans is ethically unacceptable, the opinions are much more diverse regarding the cloning of animals. This diversity in opinions about animal cloning is also reflected in the discussions concerning other forms of animal biotechnology, and thus are part of the larger social controversy regarding biotechnology in general.

The technological challenge of animal biotechnology is, What *can* we do? This question is closely followed by the ethical question, What *ought* we to do? As with any other human activity, this question needs to be answered if we are to decide what to do. That is the blessing and the curse of being an ethical creature equipped with a conscience. We cannot just act and use the possibilities that lie before us, but have to reflect on what values should be promoted and protected through our actions. This does not mean, of course, that we necessarily will agree on what ethical concerns are relevant in relation to animal biotechnology. There are huge disagreements about what the relationship between *can do* and *ought to do* is. As a result, ethics is not something we can add as a mere afterthought to animal biotechnology or something that only opponents of the technology are concerned about. Ethical reflection has to be there from the outset.

Ethical thinking, the narrowing of possible courses of actions through reflection on values, is something every human being does. We are ethical creatures who obviously have the potential to act unethically, but we cannot choose to be a-ethical. Thus, the ethical discussion of animal biotechnology should not be seen as something intrinsically different from the practical world of animal biotechnology as if the two could take place without connection to each other. Action and reflection, technology and ethics are closely intertwined, and any attempt to separate them too much will only lead to ethical reflection out of touch with reality and technological development without regard for the ethical dimension of human existence.

This chapter will begin by looking at the ethical concerns that are typically put forth in discussions about animal biotechnology in the literature and in debates in the public sphere. The approach will be to organize these concerns into two overriding categories: risks to humans and risks to animals. We acknowledge that this systematization has its limitations. Some concerns are interconnected and can only be fully understood across the categories. This is a basic condition for all attempts to bring order to the complex reality of human actions and thinking. Clarity, understood as accessibility, often comes at the cost of complexity. It is nevertheless our hope that we have struck the right balance between giving a comprehensible overview of the field while maintaining the complexity of the ethical concerns. The final section of the chapter will present some more general thoughts about different ways of understanding the role of ethics in relation to animal biotechnology and will discuss what can reasonably be expected by engaging in ethical discussions.

The Ethical Concerns

To comprehend the diversity of the ethical issues in discussions of animal biotechnology, it can be useful to begin with one of the central questions: *Are there any ethical issues specifically related to animal biotechnology, or are they the same as the ones arising in relation to more traditional ways of using animals (e.g., selective breeding within agriculture)?* This question is almost always raised at scientific conferences on biotechnology. What seems to be implied by those who raise the question is that if the presenter cannot come up with an ethical issue that is raised by that particular technology but no other, then there is no reason to take the problem into consideration. Then, it is just the same old problems and concerns and these can be adequately dealt with elsewhere. Sometimes it is even stated that it is unfair to place so much ethical focus on animal biotechnology when the concerns there are very similar to the ones found in more traditional ways of using animals, ways that are supposedly already accepted by society.

But what if it is true that there are no ethical issues raised within animal biotechnology that cannot be found, at least to some degree, in other areas of animal use? Is it then true that we can stop concerning ourselves about them, since these issues should be dealt with elsewhere – if they have indeed not already been solved? Can we then stop caring about the ethics of animal biotechnology? Or could it be that we as a society to a large extent have been unknowing about what goes on in different kinds of animal usage such as pet-dog breeding and farm-animal breeding? Perhaps even deliberately unknowing in the way discussed by J.M. Coetzee in his provocative essay *The Lives of Animals* (Coetzee 1999) where we (Western civilization) are accused of deliberately closing our eyes to the suffering we induce in animals to reach our own goals. Could it be that animal biotechnology, being the latest and most powerful tool to control the biology of animals, instead of being redeemed because of its close connection to more traditional technologies, should be the occasion for us to reflect more critically about our use of animals in relation both to biotechnology and to other areas as well (Gjerris and Sandøe 2008)?

To be able to give a satisfying answer to this question, it will be useful to look more closely at the most prominent ethical issues discussed in the literature and in the public sphere in relation to animal biotechnology. We will do this by dividing them into two overriding categories: risks to humans and risks to animals.

Risks to Humans

The ethical concerns about how animal biotechnology will affect humans relate to direct effects to human health, indirect effects through environmental impact, effects from socio-economic impacts, and risks related to the psychological effects of the possible degradation of the relationship between animals and humans. All these concerns focus on the ethical obligations that we have towards other humans. They are, to use a common term in environmental ethics, *anthropocentric*. Another thing they have in common is that it is very difficult to evaluate what the risk actually is. As risk is typically understood as the probability of a harm multiplied by the severity of that harm, and as it can be very hard to be specific about the probability and even harder to agree about how severe the consequences might be, it should be obvious that different people evaluate the below-mentioned risks differently. And things only get more complicated when unintended consequences are included in the calculations. Or, as one of the participants in the Scientists focus group said:

> I think the scientists that I'm aware of – I mean, I don't think anyone is out to make a bad disease model, or develop an animal that's going to take over an ecosystem. I don't think that's anyone's intent. You know, of course, as a scientist I am never going to say that could never happen, because there could be unintended consequences. (Scientist V6)

Another quote from the focus groups, this time the Regulators group, takes this point even further:

> There are always knowledge gaps. There's – you know, the inherent variability of biology that contributes, certain uncertainty factors, long-term effects, the fact that in genetically modifying an animal there may be in-

advertent genetic implications that are very difficult to predict. (Regulator V1)

A thorough evaluation of the risks mentioned here is beyond the scope of this book. What we are doing here is not reporting the risks that we believe should be taken seriously in the debate, but providing an overview of the risks discussed. This should be borne in mind throughout this chapter. Furthermore, the concept of risks is discussed in several chapters in this volume; see especially those by Brunk and Hartley, Reis, and Phillips.

HUMAN HEALTH

With regard to humans, the focus is on whether animals produced by biotechnologies such as genetic modification or cloning pose a risk to human health. Is it dangerous to drink milk from genetically modified cows or eat medicine produced in the milk of cloned goats? Will the animals be carriers of new and contagious diseases or will the change in their genome in other ways be a risk for human health? Only a limited amount of research has been done in this area, but what has been done has shown no additional risks in comparison with most conventionally produced animals and products (National Research Council 2002; Gamborg et al. 2006; FDA 2008). Hardly any studies have been made on food products from GM animals, whereas medicines produced through the means of GM animals (bioreactors) have undergone the usual rigorous testing of drugs. More studies have been made on products derived from cloned animals and their progeny, and all have failed to observe, as mentioned above, any significant differences between these products and those from conventionally bred animals.

With regard to genetically modified animals, a major concern related to human health is the possibility that the genes transferred might cause allergic reactions in people. If a gene from a plant is transferred into an animal and that gene is responsible for the production of a protein that makes some people allergic to the plant, then that gene could very well make the same people allergic to the meat or milk from the animal. But as there is no way to tell whether the animal has this gene or not, since it will in all likelihood not be visible, this could cause serious problems for people who suffer from various kinds of food allergies. There are ways of avoiding the most obvious problems here by choosing genes that are not known to produce allergenic proteins or by labelling all products from GM animals. But the whole area of food allergy is scientifically complicated, and there are many unknowns. Basically, it would seem that the risks in this area can be managed, but it will take a lot of consideration for each suggested product as to which genes should be used for modifying which animals. A thorough discussion of this issue in relation to plant biotechnology can be found in a report from the Royal Society of Canada from 2001 (Royal Society of Canada 2001).

Another area of importance is whether the physical composition of the food product will change and thus possibly change its nutritional value. This result can be both intentional and unintentional. One could possibly attempt to genetically modify an animal to change the nutritional value of its meat and milk. It could be the amount or composition of fat in pigs as in the case of the omega-3 pig, used as an example in the focus groups, or the amount of lactoferrin in milk. In such a case, it would be necessary to evaluate how these changes could affect human health in comparison with traditional products. But these changes could also be unintended and be caused by changes in genes that are not directly linked to the nutritional value of an animal. Molecular biology is a very complex science that tells us that genes are interconnected and seldom 'work' alone. What happens to one gene can have effects elsewhere, as was discussed in the previous chapter. Thus, changes in genes related, for instance, to the environmental impact of an animal might have unintended side effects in that they also affect the nutritional value. Only through good testing and control procedures could such problems be met in a responsible way. But as no genetically modified animals have yet (as of this writing) been approved for the agricultural market, and only a few have been developed and tested, it is very hard to say precisely how large a risk this might be. It should be clear, however, that this aspect will need to be tested on a case-by-case basis. It will depend on the gene inserted and the animal modified, and it will, in all likelihood, not be possible to say anything general about this issue. Furthermore, it will be an area where many values will clash, as it is in the end a question about how the technology should be regulated, whether the products should be labelled, and so on. For more discussion on these topics, see the final chapter of this book.

One area where it can be foreseen that the technology might involve serious risks to humans is the case of xenotransplantation of pig organs, where viruses dormant in the pig genome could be transferred to humans in the organs and possibly be awoken by the new environment in the human host organism, and thus give rise to a new emergent disease against which humans would have no natural defences. Notorious examples of emergent diseases transferred from animals to humans are the Spanish disease, AIDS, and avian influenza. In a worst-case scenario, just one xenotransplantation 'gone wrong' could cause a pandemic (Fishman and Patience 2004). This is discussed in greater detail in chapter 2.

The degree of scientific uncertainty is rather high in this area, as is generally the case in emergent fields of research. But it should be noted that the risks related to human health caused by transgenic or cloned animals thus far seem manageable, if we exclude the notion of transplanting live tissue from one species to another (xenotransplantation).

ECONOMY, PSYCHOLOGY, AND THE ENVIRONMENT

Other concerns focus on the possible socio-economic changes that animal biotechnology could cause, especially if it is ever integrated on a large scale into farming and food production. These concerns were very clearly stated by one of the participants in the Alternative Agriculture focus group:

> One other thing I wanted to bring up was around the increased control by the agribusiness and pharmaceutical sectors of agriculture, [which] leads directly to a greater economic strain on small farmers and makes it much more difficult for small farmers. And this has implications not just here in Canada but can have implications on the global south as well where agriculture is much less technological than it is here in Canada, for example. (Alternative Agriculture stakeholder V1)

However, as the technology has not been developed, let alone introduced, in this area, it is very hard to guess what will happen. However, concerns that the technologies will strengthen the movement towards large-scale farming and could further the divide between the developed and the developing world are not unrealistic if one considers what usually happens when new technology is introduced and what has happened within the sphere of plant biotechnology (Gjerris 2006).

For a further discussion of these concerns, see the chapter by Thompson in this volume.

Another kind of concern for humans that shows up is that the continuing reification or commodification of nature, where nature is seen exclusively as a resource to be used by humans, will harden human ethical sensitivity in general, thus causing ethical problems between humans as well. This argument has been brought forth against the unethical use of animals throughout the history of Western philosophy, as for instance in the work of Thomas Aquinas (ca. 1225–74). It is closely connected to concerns that humans, by turning what is strange and unknown into biological factories, lose an important aspect of what it is to be a living being sharing the world with other living beings, thus diminishing the human life-world. Or, as it was put by one of the Health Care Providers and Patient Advocates in the focus groups: 'I guess it was the deep questions of identity. And if I think of how far we're prepared to manipulate the natural world towards our own end, that felt like it was going so far that it was at the level of just abhorrent, and I know that's not a scientific reason, but it is around questions of identity, deep questions of identity' (Health Care Providers and Patient Advocates stakeholder V2).

In addition to these concerns, it is often mentioned that there could be a development whereby, because we accept applications of biotechnology on animals, we gradually change our view on these applications and end up accepting them for use on humans. This 'slippery-slope' argument seems to have been the basis of the first wave of concern regarding animal cloning when the sheep Dolly hit the world media in 1997. Thus, the problem was not animal cloning (point A) in itself, which was seen as ethically acceptable, but human cloning (point B), which was considered unacceptable, but at the end of an inevitable slippery slope that would lead down from the acceptance of animal cloning. Thus, the argument goes, we should refrain from going to point A (although it is perfectly acceptable to do so) because it automatically leads to point B, where we definitely do not want to be.

Some of the ethical concerns related to animal biotechnology focus particularly on the environment. One concern here is that the new animals can escape and breed with wild populations, thus spreading their genes in an uncontrollable environment. The most cited example here are transgenic fish, for example, salmon with genetic alterations that allow for faster growth. For one participant from the Scientists focus group, this was even the largest concern:

> When you're engineering aquatic organisms or insects, the notion is that you're applying them in an open environment, and it's not a pig or chicken, or a cow, you can contain very effectively. And so there are real issues in terms of environmental – possible environmental effects. It could come from inter-breeding with wild populations or simply from ecological interactions for – so that's my largest concern. (Scientist V3)

The concerns in this area can be about either the indirect consequences for humans, in this case economic losses for the fishing industry, if such a transgenic fish could cause havoc to the wild-living species that are already under pressure from intense fishing, or direct concerns for the animals and the wider ecosystem (Royal Society of Canada 2001; Pew Initiative on Food and Biotechnology 2003). With the exception of fish, individual animals produced by biotechnological methods are rather easy to confine, which makes it less likely that they will be a hazard for the environment. But here, as with most other concerns discussed so far, it will be necessary to evaluate the animals case by case to estimate the risks: in this instance by assessing how the animals might possibly interact with the environment and whether this will be a threat to human interests.

Another concern, one of respect for nature, represents the view that changes in nature brought about by humans are unwanted, even if they do not pose a risk to humans. There is a feeling that the technologies are changing things that should not be changed and touched. Basically, there exists an appreciation of untouched and un-interfered-with nature. And although hardly anything on the planet can avoid some influence by humans, biotechnology seems to some as a very direct and brutal way of mingling with the environment. A participant in the Alternative Agriculture group voiced this concern in a very general way: 'Well I just – I just feel like it's just going against nature. It's unnatural processes that are happening and, you know, of course it's not so much that I am against progress and how – but it's how we are getting there. That is the – a very large concern' (Alternative Agriculture stakeholder V2).

We will return to this problem at the end of the next section.

Risks to Animals

The other large group of ethical issues has to do with animal welfare. There seem to be basically two kinds of concerns at stake in the debate.

The first focuses on keeping animals healthy and avoiding pain and other kinds of suffering, and on promoting positive experiences: in general, this conception focuses on the health and subjective experiences of animals and entails a high degree of identification with them. One of the participants from the Agricultural Producers focus group phrased it this way while discussing the cancer mouse example presented to the group:

> Oh, I don't know. I mean I support using animals for meat but just making animals sick and watching them go through that and – anybody who thinks that farmers don't care for their animals, they don't know what they are talking about. Because probably the most offensive thing for me in the last twenty years was watching my pigs go through circovirus. It was just – you go in the house and cry because there's nothing you can do to make them feel better, and here we are intentionally making a mouse sick. (Agricultural Producer V4)

Besides these considerations, there is another and somewhat broader perspective that includes the animal's opportunity to engage in essential species-specific kinds of behaviour. It can, of course, be debated what species-specific behaviour is and whether traits that are bred into animals can be said to be species-specific. We will discuss this later. For now it is enough to refer to the almost intuitive knowledge that most of us have about what an animal should do. The American ethicist Bernard Rollin has expressed it this way: 'Animals, too, have natures – the pigness of the pig, the cowness of the cow, "fish gotta swim, birds gotta fly" – which are as essential to their wellbeing as speech and assembly are to us. Common sense tells us that animals who are built to move need to move to feel good, there is no point in trying to prove that they are fine if kept immobile – no one will believe you anyway' (Rollin 1995: 159).

Both concerns focusing on the subjective experience of the animal and those focusing on the broader notion of animal welfare are widespread in Western societies today. This was exemplified in the focus groups, where these concerns repeatedly turned up. In the following we will examine how these concerns relate to more specific uses of animal biotechnology.

Biotechnology has been applied to animals mainly within biological research and to produce disease models. Typically, the goal of modifying the animal is to produce animals that either under- or over-express certain genes, or that express a mutated, disease-causing human gene. In all these cases, functions in the animal's body are in some way disrupted. In principle, modifications can involve any part of the animal genome, and the effects on the animal's phenotype range from those that are lethal to those that have no detectable effect on the health of the animal. It is therefore impossible to generalize about the welfare effects of the genetic modification of animals (Olsson and Sandøe 2004).

However, effects may be divided into two main categories: the intended and the unintended. Welfare problems related to intended genetic change are hard to avoid, since the very point of inducing the change is to affect the animal. Thus, all mice carrying the human Huntington's disease gene will suffer welfare problems as a consequence, including rapid progressive loss of neural control leading to premature death if they live long enough to develop the disease. Unintended effects are connected with the present inaccuracy of the technology and our insufficient understanding of the function of different genes in different organisms. Both of these factors are responsible for the rather unpredictable nature of genetic modification at the phenotypic level.

However, it is likely that at least some of the unintended welfare problems can be avoided as the technology and our scientific understanding develop. Regarding both unintended and intended consequences of genetic modification (for example, in creating a disease model), it may be possible to predict welfare consequences using information about the effects of similar modifications in other species, including human disease symptoms. This could enable the producers of the animal to consider at least some of these consequences before the animal is actually produced (Dahl et al. 2003).

One example that can clarify the complex welfare problems within animal biotechnology is animal cloning. Success rates so far have been low (0, 1–20 per cent), and of the few individuals born, many suffer from impaired health and welfare. Problems include placental abnormalities, foetal overgrowth, prolonged gestation, stillbirth, respiratory failure and circulatory problems, malformations in the urogenital tract, malformations in the liver and brain, immune dysfunction, anaemia, bacterial and viral infections, and so on. Some of these conditions are gathered under the term Large Offspring Syndrome (LOS). LOS is often seen in cloned animals, especially cattle, but it also occurs in cases where animals are conceived by means of in vitro fertilization.

It is not yet clear whether the welfare problems experienced by cloned animals can be avoided through technological or methodological improvements or whether there are deeper epigenetic factors behind them. What is clear is that the technology seems to cause massive welfare problems as the animals are produced, whereas the animals that are successful clones do not suffer more than non-cloned animals (Vajta and Gjerris 2006).

If we look at animal cloning through the narrower perspective of animal welfare, it is obvious that some of the problems mentioned give rise to concerns. All animal suffering from the point where the animal is able to experience it is ethically problematic. However, the loss of insentient foetuses is not problematic unless it places another sentient animal (such as the surrogate mother) under strain. But from the broader perspective of animal welfare, this loss of animal life is also important, just as is the question about the extent to which the animal is allowed to fulfil its species-specific potential, regardless of its subjective experience. The broader perspective can perhaps best be understood as pointing to an additional group of considerations that has to be taken into account when assessing animal welfare. It should be remembered that being concerned with the opportunity of the animal to engage in certain kinds of behaviour does not exclude caring about the subjective experiences of the animal. Nevertheless, sometimes these two viewpoints clash; in that situation, it becomes important to clarify what kind of perspective is in play. Considerations within the narrow perspective regarding the subjective experiences of the animal might be outweighed by the other considerations included in the broader perspective.

An illustrative dilemma concerns the evaluation of the welfare of battery hens and free-range hens. Although biotechnology is not involved here, the case serves perfectly to point out the important differences. From a narrow perspective, it is not ethically problematic to deny the animal the possibility to follow its instincts (as battery-cage egg production does) as long as this does not affect the subjective welfare of the animal (lead to negative experiences). Although one can rarely prevent an animal from following its instincts without causing suffering, still through breeding (conventional breeding or breeding involving cloning or genetic engineering) changes could in theory be made in the animal that would make it more fit for the conditions under which it will have to live. This would not lead to any negative subjective consequences for the individual hens once they were 'working,' and such a use of biotechnology would be seen as ethically unproblematic from the narrow perspective (Gjerris et al. 2006). Some of the welfare problems caused by battery-cage egg production could therefore in theory be solved through breeding hens that were adapted to the conditions of battery-cage production rather than changing the conditions (Rollin 1995).

There are certain practical problems that make this scenario unlikely in the foreseeable future. First, the traits to be bred out of the animal will have a complex genetic background, since the objective is to produce an animal in which all motivations other than those that can be satisfied in a battery cage are eradicated. Second, it will be a difficult challenge to ensure that one is indeed breeding for an animal with a restricted set of motivations rather than an animal that reacts passively, or even with apathy, to adverse conditions. This is not to say that breeding for behavioural traits cannot be used to improve animal welfare (problem behaviours such as feather pecking in hens have indeed been shown to be under genetic control), only that the production of such an animal would face huge challenges.

From the broader perspective, the very idea that we should breed hens to cope with battery cages raises serious worries. Instead of changing the hens, it would be ethically preferable to change the production systems in a way that would allow the hens to fulfil their natural potential. Here it needs to be said that life as a free-range hen is in some ways less protected than life as a battery hen. Disease, feather pecking, and cannibalism occur frequently within flocks of hens. It is part of the species-specific behaviour of hens when kept in large flocks. Nevertheless, from a broader perspective, this may be an acceptable situation, since it is counterbalanced by the fact that the hens are living more naturally (Gjerris et al. 2006).

Another concern within the broader perspective is that many of the new technologies extend control over procreation. This control is already widespread within animal breeding through the use of semen collection, artificial insemination, superovulation, embryo transfer, transvaginal ovum pickup, and so on. The question is whether this interference is ethically acceptable, since all the technologies mentioned can, in very general terms, be described as unnatural when compared to the 'natural' life of animals. However, the very idea of naturalness as something valuable in itself raises questions about how naturalness should be understood. From animals used in basic research to farm animals bred for production, one can question if anything in their lives is natural. We will return to this point shortly. For now the question can be rephrased to focus on the extent to which the domesticated animals have an opportunity to fulfil their species-specific behaviour within the framework of the domestication process. Thus, a laboratory mouse will live its life in a cage, but it might nonetheless fulfil certain species-specific behaviour (for example, digging or nest building) if given the chance.

At this point it is necessary to distinguish between two different viewpoints within the broader animal welfare perspective: one where respect is aimed at the original genotype, and where deliberate changes to animals are seen as inherently problematic, and one where respect is aimed at whatever animals come out of the process of breeding, and where the potential problems concern the development of 'un-harmonic' animals. To people taking the first viewpoint, the notion of deliberately breeding hens that have such limited potential as to be content with life as a battery-cage hen is seen as ethically problematic in a way that might outweigh the advantages of the lowering of actual animal suffering. It is as if something just is missed in the human – animal relationship when animals with less potential than normal are deliberately produced. Inherent in this version of the broader perspective is a certain respect for the natural state of the animal. Although it is intuitively compelling to many people, it has to be pointed out that this perspective suffers from an inherent ambiguity when domesticated animals are discussed, since it is almost impossible to point to a stage in the development of such animals that would constitute their 'natural' state and thus be the developmental point that should be respected (Gjerris et al. 2006).

The ambiguity of the concept of biological naturalness is a leading reason why other thinkers have suggested a different way of considering animal welfare problems within the broader perspective. They believe that the natural behaviour of the animal is to be respected, but that natural behaviour is not seen as something static. And just as domesticated animals have been bred to be better adapted to housing in confinement in the past, through the selection of individuals that were most productive in a specific environment, animals today can be bred, either conventionally or through genetic modification, to be better adapted for modern-day production systems. What is to be respected is the actual nature of the new animal and not some past ideal (Rollin 1995).

Whether we look at animal welfare from a narrow perspective or one of the broader perspectives, two important issues must be borne in mind when assessing the ethical dimensions of animal biotechnology. First of all, it is important to note that, as far as consequences for the animals are concerned, the difference between traditional breeding technologies and the new biotechnological tools seems more to concern the level of welfare problems for the animals than the kinds of problems occurring. We will return to this issue later.

Another issue besides animal welfare that needs to be discussed at this stage, however, is that of animal *integrity*. Animal integrity can best be understood as an inherent limit in the relationship between humans and animals – a limit governing what it is ethically acceptable for humans to do to animals. Very often the notion of integrity shows up when technical possibilities that will change the everyday experience of the animal are discussed, as in the cases of the cancer mouse, the GM salmon, and the GloFish, which were discussed as examples of animal biotechnology in the focus groups. But the notion of integrity covers more than just what is visible to the eye. Also in the cases of the omega-3 pig, the insulin-producing goat, and the spider-silk-producing goats, it can be argued that the integrity of the animal has been violated.

In other words, integrity is a limit based on an understanding or experience of animals as beings surrounded by an invisible border that may not be transgressed. If we cross that border, we immediately encounter the integrity of the animal – the wholeness, fullness, and 'completeness' of the living being that we are now attempting to change to better fit our needs. This does not mean that we cannot and do not seek to justify our transgressions of animal integrity. Rather, it means that even if the reasons for doing so are good from an ethical perspective, we end up not necessarily making the right choice, but perhaps more choosing the lesser evil. Animal biotechnology is thus inherently ethically problematic from within this perspective (Gjerris, 2006).

One way to exemplify what is meant when it is claimed that something transgresses the integrity of an animal is to attempt to capture the distinction between the knowledge of the animal that is expressed through our understanding of its usefulness to humans and the knowledge that is expressed in our immediate experience of the animal. A cow is a producer of hide, milk, and meat; it holds no surprises when experienced from a human perspective, where the fulfilment of human need is central. But from another perspective, where the cow is understood as something independent of humans – as a life form with its own needs, history, and importance – it becomes clear that we do not know all there is to know about cows just because we know how to use them. There is something more to cows: something that in a sense alienates them from us and that should prevent us from reducing them to being merely means to our ends.

It should be clear that the idea of animal integrity both broadens the concept of animal welfare beyond the narrow perspective and rejects the notion that the naturalness of an animal is something that should be respected only in the individual animal – thus permitting humans to change the nature of animals in general, as was the case in the second version of the broader perspective. The notion of integrity falls close to the ideas of naturalness discussed in relation to the first version of the broader perspective on animal welfare. Often the term *naturalness* is even used in a way that makes it interchangeable with the term *integrity*. They both attempt to capture something essential in the experience of the animal, something that tells us the animal is worthy of ethical consideration whether it is subjectively hurt or not.

As mentioned earlier, the term 'naturalness' is quite ambiguous. First of all, it can be very hard to distinguish between what is natural and what is unnatural, and second, the term begs the question why the 'natural' automatically should more ethical than the 'unnatural.' Is a dog or a cow natural? Both are the result of thousands of years of breeding. How can anything we do to either one now be 'unnatural' in any significant way? If the natural has something to do with being independent of humans, it can be hard to see how changing just a few more genes in species that are already so thoroughly bred that they are almost unrecognizable from their ancestors can make them more 'unnatural.' And what is so ethically good about being 'natural'? There are lots of things in nature, such as earthquakes, starvation, and killing of the young, that we will have a hard time claiming are ethical simply because they happen 'naturally.' It is, it seems, rather easy to dismiss notions of unnaturalness as a valid ethical concern.

Nevertheless, the idea that there is an ethically relevant distinction to be made between the 'unnatural' and the 'natural' prevails. For some it takes a religious shape. They argue that we should not be 'playing God' or changing the 'order of creation.' To others it is a more philosophical distinction resting on the notion of a need for respect for the natural order, the forces of evolution, and so on. A series of words can be attached to the idea of naturalness that can help explain why, for many people, it is a preferable state to be in rather than being unnatural: controllable, familiar, in tune with the rest of nature, understandable, respectful, sharing, and so forth. This is not the place to go further into this discussion. We would just like to point to the fact that it is not unproblematic to use a term like 'naturalness' in discussions on animal biotechnology when it is, by all means, not self-evident what it entails. On the other hand, the term obviously is more complex than it is sometimes given credit for. So one should not dismiss it too easily.

The concepts of pain and suffering are tangibly present in the world. We can relate to them and empathize with an animal on a very basic corporeal level. To say it is wrong to cause an animal to suffer speaks to a bodily founded understanding of the world. The concepts of integrity and naturalness also speak to basic experiences of being in the world, albeit in another, less tangible manner. But very few among us are in doubt as to what is meant when someone tells us that a friend of ours behaved unnaturally or that dressing up elephants to perform tricks in a circus transgresses their integrity. Integrity and naturalness are thus very distinct concepts, although they have a wide meaning – they encompass a lot of experience.

Nevertheless, there is a certain tendency to leave these notions behind in the ethical debate about animal biotechnology. It is as if the complexity and wideness of the concepts make them less suitable for discussion. It is even sometimes stated that these are metaphysical and religious concepts – evidently in contrast to more down-to-earth and sensible concerns about risks to human health and animal welfare. This distinction can, however, be challenged, since it is not clear why the concerns for human health and animal welfare at their foundation are any more or any less founded in metaphysical assumptions about the world than the concerns for naturalness and integrity (Gjerris and Sandøe 2006). Whatever one thinks of the concepts of naturalness and integrity, it is important to realize that they play a large role in the perception of animal biotechnology for many people. It is clear from a number of European surveys that concepts such as integrity and naturalness play a significant and growing role in the general perception of the legitimate use of biotechnology on animals (Lassen et al. 2006). They are therefore worth taking seriously, even if one does not share them – at least if one seeks to understand what is going on in the ethical debate.

The notion of 'something being there' that should be protected, even though it transcends the suffering of the individual animal, was put into words by one of the participants in the Alternative Agriculture focus group when the GloFish example was discussed:

> There is this – it's almost just a philosophy of respect, and that aquarium fish example for me crossed a very strong line of respect where I felt like it's just anything that's done for a frivolous need, even if the welfare of those individual animals wasn't infringed upon at all, even if they were totally healthy, happy fish I still wouldn't support it because it seems disrespectful to that animal to tinker with its nature or its telos to actually have just a frivolous benefit. (Alternative Agriculture stakeholder V4)

The range of ethical concerns focusing on animals in relation to animal biotechnology thus extends from concerns related to the possible experiences of pain and suffering in the individual animal to concerns about the possible violation of the integrity and naturalness of animals in general through the use of biotechnology.

Is Animal Biotechnology Radically New?

Having thus described the ethical issues typically discussed in relation to animal biotechnology, we now return to the question whether animal biotechnology raises any specific concerns or whether the concerns discussed above can also be found in relation to other kinds of animal usage. Those who are critical of cloning and other forms of animal biotechnology seem to imply that these technologies are problematic because they give rise to new kinds of problems (Midgley 2000; Chapman 2005). Defenders of animal biotechnology disagree and claim that there is nothing new under the sun. We continue to change animals to suit our own needs. Only the precision and effectiveness of the methods has changed (Kues and Niemann 2004). Hence, animal biotechnology raises no unique ethical problems.

It would seem that the argument made by supporters of animal biotechnology is correct in the sense that most of the welfare problems associated with cloning and genetic modification are present within more conventional technologies too. Large Offspring Syndrome is a problem not only within cloning, but also when other kinds of breeding technologies are used. The welfare problems that may arise from depriving animals of their natural procreative activity are linked to other technologies as well. Welfare problems may also occur as a consequence of selective breeding programs, as for instance when an excessively narrow focus on productivity leads to leg disorders in broiler chickens, or to increased levels of mastitis in cows.

Finally, the way that cloning and genetic modification can be said to violate the integrity of animals seems to be in clear continuation with already established practices that also treat animals only as resources for human consumption.

Nonetheless, one should be careful not to dismiss ethical concerns related to animal biotechnology merely by pointing to the similarities between earlier and new uses of animal technology. The basic assumption underlying this argument is that people have accepted the older techniques – and that might not hold true. Most of us are largely unaware of the consequences of selective breeding, and in general many of us are not comfortable with intensive confined housing systems, but in reality we were consulted on neither of these matters. The argument could therefore just as well be reversed: public worries about new biotechnologies, and the genuine ethical concerns into which they can be translated, could be seen as a reason to critically analyse not only new biotechnologies but also existing ones, and be a trigger for renewed discussions about what it is ethically acceptable to do to animals. Thus, while animal biotechnology might not be something radically new, it could nonetheless be the straw that breaks the camel's back (Cooper 1998).

What Can Ethics Do?

Generally there is no disagreement between the different stakeholders about the importance of a qualified discussion about the ethical aspects of animal biotechnology. But the agreement ends there. What is to be discussed, how it is to be discussed, with whom it is to be discussed, and the purpose of the discussion are matters of strong disagreement. But the matter runs deeper than that, because when one sifts through the literature, it becomes obvious that there is not even agreement about what 'ethics' is. The discussion of ethics is too encompassing to cover here in detail. But if we simplify the discussion, ethics is usually understood in two ways. To some ethics is as a problem-solver – a hammer, so to speak – that can be used to solve the problems of consciousness that might arise when developing new technologies. Thus, research projects often have a small amount of funds allocated to delve into ethical and social-scientific questions. The hope is that this will somehow 'solve' the ethical concerns about the research or prevent the research from raising concerns in the first place.

The other way to see ethics is as a flashlight, as a way of reflecting on matters to clarify them, but probably also to complicate them, but distinctively not to solve anything. Ethical concerns seldom disappear just because one reflects upon them. Nonetheless, it is crucial to have ethical reflection to grasp just how complicated the choices are that we have to make. The idea that identifying the problems is the basic task of ethics and that this is helpful, albeit in an indirect way, for the choices we make, is the underlying assumption of this book. But why is ethics not a solution in itself? To understand this, it can be helpful to look at the history of modern applied ethics.

The discipline of medical ethics arose as an answer to the atrocities in the concentration camps of the Second World War. This was the first time that ethical analysis was used as an academic tool to examine ethical questions related to a specific area of human actions. And ethics went from academic oblivion into governmental committees, editorials, and public meetings to such a degree that the British philosopher Stephen Toulmin in the mid-1970s could write an article entitled 'How Medicine Saved the Life of Ethics' (Toulmin 1982).

The next area to get its own brand of ethics was the environment. Environmental ethics followed closely in the footsteps of the dawning realization of the ecological crisis in the 1960s. Soon after came bioethics as a devoted follower of the technological developments within medicine and agriculture. In the meantime, the world has also seen the advent of nursing ethics, business ethics, legal ethics, and so on. It is evident that ethics has become popular. It has become a good companion whether you are a private person, a politician, or a company. It simply pays to be 'ethical.' It is a kind of social monetary unit that buys acceptance. At the same time, ethics has been specialized to such an extreme that almost all sectors of human behaviour are seen to demand their own kind of ethics.

This development entails many problems – especially because it places ethics in a situation where it must inevitably disappoint many people or cease to be true to its own limitations. And it has created a climate for ethical debates where it is becoming more and more difficult to ask the general ethical questions that most of our actions raise, whether or not they fall in the area of one or another of the applied ethics. The first problem is illustrated by the fact that although the ethical aspects of genetically modified organisms have been discussed for the past twenty-five years in the Western world, no consensus whatsoever has been reached. It is as if the participants in the debate have just dug the trenches deeper. Obviously, it is not just that the ethical problems, concerns, and aspects of a technology disappear because they are discussed. Rather, what seems to happen is that the participants in the discussion become more and more convinced about the strength of their own arguments.

We all live our lives governed by values. Those values can be very different from, even to the point of contradicting, each other. But they share the basic condition of all values. None of them are self-evident. They cannot be proved or defended in a rational technical way. There is simply no logical way to decide whether socialism or liberalism, utilitarianism or deontology, or any other basic ethical theory is the 'best' theory. While it is easy enough to decide that it is right to do *good*, it is impossible to agree upon what *good* actually is. So discussing ethics will not make us agree on what values we should hold. This is simply because ethics is not about facts (although they can play a role in the debate), but primarily about values, which can lead to disagreement on what kind of life and society we should strive to create.

What then is the idea of doing ethics? Seen from this perspective, what we can reasonably expect ethics to do for us is to help us clarify our own values and gain a deeper understanding of the values of others. This understanding does not necessarily lead to any kind of agreement, but it might, with a bit of patience, lead to a greater respect for the values of others. The Danish philosopher Ulli Zeitler has phrased it this way: 'Essentially, the task and reasonable expectation of philosophical activity is not to solve problems, although we may advance considerably by clearing up the central concepts, but opening our eyes to previously unconsidered problems. The last function is crucial for giving new directions to future inquiries' (Zeitler 1997, p. 39).

The other problem that the development within modern ethics has made blatantly clear is that the tendency to treat each area of human activity as ethically distinct from all others and therefore in need of its very own 'ethics' comes at a price. It is obviously not wrong in any way to inquire into a given human practice to see if it entails some specific ethical concerns. The problem is the tendency to disregard the need to discuss the ethical concerns that move across the board and are relevant within most, if not all, areas. Ethics is becoming a discipline where only the specific problems are worth discussing. The general problems are recognized, but it is somehow never the right time to discuss them. It is as if that which belongs to all ends up belonging to none.

A good example of this tendency is the discussion about the novelty of the ethical concerns raised by animal biotechnology, which has been noted previously in this chapter. A claim often made within the ethical discussion of animal biotechnology is that, since the animal welfare problems encountered with regard to cloning and genetically modifying animals are similar to the ones encountered within other and more familiar kinds of animal husbandry, there is no particular need to discuss these problems in the context of animal biotechnology. But where are we to discuss them then? There are no fora for ethical discussions as such. If they are to be fruitful in any way, ethical discussions are strongly in need of a context, of being discussed within the area from which they arise. Otherwise, ethical discussion will just be an academic sideshow of grumpy old men mumbling and scratching their beards, and tense younger women with stern glasses and earnest voices who keep claiming that 'they told us so.'

Conclusions

Is there a 'right' way to do ethics? It is doubtful. The best one can do is to reflect upon ethics in a continued awareness of the perspective from which one sees the world. There is no view *from nowhere*, as the philosopher Thomas Nagel has said, where we can go to gain an objective insight into the ethical aspects of new technologies. What we can do is simply be honest about the views that govern our sight and thereby decide what aspects of a situation we are open to and believe that these are important aspects. And then we can try to listen to other perspectives, not necessarily agreeing with them, but at least taking their thoughts into consideration. Again, this might not solve any problems, but nobody ever promised us that ethics could do that.

This view of how ethics can best be conducted in a fruitful manner has been the core idea behind this book. We have invited different stakeholders in the area of animal biotechnology to share their perspectives with us through the focus groups. These perspectives are the foundation of this book, which seeks to lay out the ethical landscape of animal biotechnology. We realize that it can be frustrating not to be given any clear answers on all the ethical concerns discussed in this chapter and in the rest of the book. But in order to eventually find a way through all the hopes and concerns, we believe it is important to begin by listening to those who will share their views. That is what we have tried to do in this chapter and what we will do in the rest of this book.

References

The Center for Food Safety. 2007. *Not ready for prime time: FDA's flawed approach to assessing the safety of food from animal clones*. Washington, DC: Center for Food Safety.

Chapman, A. 2005. Genetic engineering: The unnatural argument. *Techné* 9(2): 81–93.

Coetzee, J.M. 1999. *The lives of animals*. Princeton: Princeton University Press.

Cooper, D.E. 1998. Intervention, humility and animal integrity. In *Animal biotechnology and ethics*, ed. A. Holland and A. Johnson, 145–55. London: Chapman and Hall.

Dahl, K., Sandøe, P., Johnsen, P.F., Lassen, J., and Kornerup Hansen, A. 2003. Outline of a risk assessment: The welfare of future xeno-donor pigs. *Animal Welfare* 12: 219–37.

Fishman, J.A., and Patience, C. 2004. Xenotransplantation: infectious risks revisited. *American Journal of Transplantation* 4: 1383–90.

Gamborg, C., Gjerris, M., Gunning, J., Hartlev, M., Meyer, G., Sandøe, P., and Tveit, G. 2006. *Regulating farm animal cloning. Recommendations from the project Cloning in Public*. Danish Centre for Bioethics and Risk Assessment.

Gjerris, M. 2006. *Ethics and farm animal cloning. An examination of the risks, values and conflicts related to farm animal cloning*. Danish Centre for Bioethics and Risk Assessment.

Gjerris, M., Olsson, A., and Sandøe, P. 2006. Animal biotechnology and animal welfare. In *Ethical eye – Animal welfare*, 89–110. Strasbourg Cedex: Council of Europe.

Gjerris, M., and Sandøe, P. 2006. Farm animal cloning: The role of the concept of animal integrity in debating and regulating the technology. In *Ethics and the politics of food*, ed. M. Kaiser and M.E. Lien, 320–4. Preprints of the 6th congress of the European Society for Agricultural and Food Ethics. Wageningen, The Netherlands: Wageningen Academic Publishers.

– 2008. Transgenic animals. In *Encyclopedia of environmental ethics and philosophy*, ed. J.B. Callicott and R. Frodeman. Detroit: Macmillan Reference USA.

Kues, W.A., and Niemann, H. 2004. The contribution of farm animals to human health. *Trends in Biotechnology* 22(6): 286–94.

Lassen, J., Gjerris, M., and Sandøe, P. 2006. After Dolly: Ethical limits to the use of biotechnology on farm animals. *Theriogenology* 65(5): 992–1004.

Midgley, M. 2000. Biotechnology and monstrosity: Why we should pay attention to the 'yuk factor.' *Hastings Center Report* 30(5): 7–15.

National Research Council of the National Academies. 2002. *Animal biotechnology: Science-based concerns*. Washington, DC: National Academy of Sciences.

Olsson, I.A.S., and Sandøe, P. 2004. Ethical decisions concerning animal biotechnology: What is the role of animal welfare science? *Animal Welfare* 13: 139–44.

Rollin, B.E. 1995. *Frankenstein syndrome: Ethical and social issues in the genetic engineering of animals*. Cambridge: Cambridge University Press.

Royal Society of Canada. 2001. *Elements of precaution: Recommendations for the regulation of food biotechnology in Canada*. Expert panel report on the future of food biotechnology. Ottawa: Royal Society of Canada.

Toulmin, S. 1982. How medicine saved the life of ethics. *Perspectives in Biology and Medicine* 25(4): 736–50.

United States Food and Drug Administration. 2008. *Animal cloning: A risk assessment*. Washington, DC: US Food and Drug Administration.

Vajta, G., and Gjerris, M. 2006. Science and technology of farm animal cloning: State of the art. *Animal Reproduction Science* 92(3–4): 210–30.

Zeitler, U. 1997. Transport ethics: An ethical analysis of the impact of passenger transport on human and non-human nature. Centre for Social Science Research on the Environment, University of Århus, Denmark.

4 The Biotechnology Industry and Agricultural Economics

PETER W.B. PHILLIPS

The global agri-food system has gone through a rapid technological and industrial transformation in the past generation, and faces even greater challenges with the experimental use of transgenic technologies and planned commercial introduction of genetically modified animals. While it might be tempting to examine the ethical perspectives of this event in isolation from the broader context, it has meaning only when considered in relation to the overarching political economy of the industry and other significant trends in the production of protein through animals.

Ultimately, the global agri-food industry operates under the assumption that the fundamental issue in the twenty-first century will be to efficiently, effectively, sustainably, equitably, and ethically produce, access, and use protein, fibre, micronutrients, and inorganic materials to power our bodies and our society. This can be observed in vision and mission statements from a range of actors, including Monsanto, the Food and Agriculture Organization, and the Canadian Federation of Agriculture. They acknowledge that this challenge affects (or should affect) a wide array of our natural and human systems.

Animal production is embedded in a large set of interlinked systems of agri-food and industrial production. Some societies continue to live in a pastoral way with native populations of animals they harvest for food, but most of the animal proteins we consume are the product of long-standing and intensive management of animals and their genetics, diet, and environment. Many of the recent transformations in the global agri-food system have profoundly affected for most of us all aspects of who, what, where, when, why, and how we breed, raise, and use animals.

Any consideration of new technology must necessarily be global, or at least be globally aware. While Canada may have a unique and self-contained set of issues and processes in some areas, most of the Canadian scientists, biotechnology and genetics companies, producer organizations, meat processors, wholesale and retail companies, regulators, consumers' associations, faith groups, and NGOs operate at the global level, or are at least interconnected or networked internationally. They thereby define norms, standards, conventions, values, and processes that determine the acceptability of new technologies.

The production of animals is highly value laden. While profit is the first and most important goal of industry, many producers have strong moral codes that place them in the position of custodians and protectors of their animals and the environment. Although many consumers see food simply as fuel, a significant number of people and groups in society invest food with cultural or religious significance. Moreover, while the industrial supply chain – stretching from test tube to dinner plate – is often perceived as amoral and solely profit driven, the reality is often different. Firms, whether owner-operated farms or large multinational corporations, are composed of people who have moral and ethical perspectives which, at times, are reflected in internal debates or corporate actions. While in the past these perspectives were viewed as personal and had to be divined in the context of commercial activity, many firms and industries are moving towards corporate social responsibility (CSR), sometimes called the triple bottom line, by formally considering and incorporating into the management of their organizations, in order of priority, profits, the environment, and social responsibility. While one might question the underlying motivation for CSR, it is in some modest ways beginning to be reflected in the actions of some firms.

In this context, the proposed introduction of transgenics in and around the animal food chain has raised fundamental questions about both the new applications and the pre-existing structure and function of animal production. This chapter offers a discussion of the ethical dimensions that are implicitly or explicitly considered in the application of transgenics and any other new technology into the industrial agri-food system (in contrast to most other chapter authors in this book, I will not consider cloning technologies explicitly). It will review the context of the new technology, examine the possible impacts of adjacent possible applications of this technology in the animal industry, and provide an assessment of the gaps in the ethical consideration of new technologies. At the same time, the chapter draws on the data generated from the focus group research that underpins this book.

Ethical Processes

Economic and commercial activity appears to many to be an amoral system that is driven solely by profit. This is at least partly due to the reality that farms and firms cannot be self-sustaining in the long run without generating revenues that exceed their expenses. While a cursory overview of the activities and outcomes of the commercial and economic system could lead to the conclusion that it is bereft of a moral compass, the reality is both more complex and dynamic.

Economics and commerce as we know them today are products of the Enlightenment and the first industrial revolution (roughly 1750–1850). A seminal event was the publication in 1776 of *An Inquiry into the Nature and Causes of the Wealth of Nations* by Adam Smith, chair of moral philosophy at the University of Glasgow. This revolutionary document laid out the underpinnings of a liberal, individualistic-based economy, which ultimately helped to redraw the boundary between the state, the individual, and the market. John Locke had earlier posited that citizens should be entitled to the fruit of their labours, establishing the notion of the individual as an independent source of wealth, power, and authority (Locke 1690). Smith's work extended this notion of the individual as separate from the state and other authorities. He did this quite deliberately; his supporters, promoters, and admirers similarly had an overt, political objective in promoting the concept of self-regulating markets as a way to limit the power of the state (Gilpin 2001). Before this, the feudal system dominated in many regions, with individuals having little or no independent role as producers or consumers. While there was some entrepreneurial development previously, the political system continued to constrain the role of the merchant class. By proposing that self-interested individuals could promote both social and economic improvement, Smith provided a key underpinning to the emerging liberal conception of government based on a social contract.

Essentially, economic theory generally posits that individuals as producers, left to their own devices, will allocate their labour (either directly in their own production or through selling their time to others for a wage) up to the point where the 'pain' inflicted by the last unit of work equals the 'pleasure' that could be derived from the output or wages one generates with that income. Similarly, individuals as consumers will buy goods or services as long as the 'pleasure' (i.e., utility) generated is greater than or equal to the cost of buying those goods (and the related 'pain' of the work required to earn that income). Individuals following their personal interest would thereby allocate their own labour and income. Theoreticians have demonstrated that it is possible that an economy made up of such self-satisfying individuals could produce and consume an 'optimal' amount of economic activity. This optimum is defined as the point at which maximum utility is reached – working longer hours to consume more would be a loss in overall utility because the pain of the last unit of work would exceed the pleasure of the last unit of goods consumed, while working fewer hours would lower the pain but would lead to disproportionately larger declines in pleasure from consuming goods.

Economics is thus a moral philosophy – albeit a limited one. Given its underlying assumptions, it unapologetically divides state and market and disenfranchises those without resources (either labour or wealth) from access to the system. In the process, the choices about what will be produced and consumed in aggregate are beyond the control of any individual actor – be it government, industry, or any single individual.

The optimal outcome, called 'perfect competition,' is more of a theoretical construct than an observable reality. Theory shows that it can only be realized if five rigorous and highly unlikely conditions hold: there must be a large number of producers and consumers in a given market, each so small that their individual actions have no significant impact on others; the goods and services must be perfectly substitutable; there must be full information on the options; there must be no barriers to entry or exit for either producers or consumers; and there must be no external effects of any production or consumption that lie outside the market. As different sets of assumptions are relaxed or break down, different types of imperfections arise. Monopoly and oligopoly, where one seller or a small number of sellers respectively exert market power, occur with barriers to entry (e.g., large fixed costs to enter) or imperfect information (e.g., patents). Markets are likely to fail to produce optimal quantities of 'public' goods (e.g., non-excludable services such as public health and safety). Moreover, even if a competitive marketplace might equalize private social benefits and costs, externalities such as pollution could still create a market failure.

Economic theory and practice suggest governments are justified in intervening in the economy when the market exhibits 'failure.' The difficulty comes in judging where and how to engage. The earliest and most morally consistent decision rule is the strict Pareto criterion, which asserts that the state should intervene only if it can make someone better off without making anyone else worse off (this is a type of economic Hippocratic oath to 'do no harm'). Given the nature of markets, however, there is virtually no intervention that will benefit some without disadvantaging others. Hence, moral philosophers have offered alternative criteria. Economists John Hicks and Nicholas Kaldor offered the notion of a compensated Pareto outcome, where in a democracy it would be acceptable if a state acts whenever it *could* compensate losers and still have winners; compensation would not need to be actually forthcoming. While the application of this rule is now used to justify a wide range of government interventions into the marketplace, it is often contested. The most common concern is that it does not take into account income distribution; there will inevitably be losers. Moreover, it can adequately address only concerns that can be priced; fundamental moral concerns are thereby often left out of the framework.

This perspective was fully reflected in the Agricultural Producers focus group. One participant asserted:

> I think government has got a responsibilty to protect human health and enviromental health. Leave the rest to us. Get out of our way . . . If you try to quantify the other stuff, that's called marketing. We read the market, we tailor what we are producing, we'll provide a stream and we'll sell it. (Agricultural Producer V4)

But others were less sanguine. A participant from the Animal Justice focus group stated that

> life has evolved over millennia, and I think the species have – as some people have said, the pigness about the pig, and the cowness about a cow. I mean, they're creatures that have . . . well developed telos, and we're messing with that. And at such a high cost. (Animal Justice stakeholder V1)

Recently, many non-state actors – both non-governmental organizations (NGOs) and firms themselves – have acknowledged the potential for new criteria for decision making. Many values-based civil authorities have recently pressed both governments and industry to internalize into their organizational or corporate goals a set of guiding principles and values and to report on related activities and performance. This effort has been variously labelled as 'the third way' in government and 'corporate social responsibility' (CSR) in industry. This movement reflects unease with monopoly state action or unbridled markets. On the corporate side, the concept of CSR has involved companies committing to pursue, measure, and report to stakeholders and society on profits, environmental protection, and social justice. In support of this effort, the Organisation for Economic Co-operation and Development (OECD) has developed a set of guidelines for multinational enterprises in an effort to spur 'responsible corporate behavior worldwide,' laying out a set of concepts, principles, and guidelines related to disclosure, employment, industrial relations, the environment, bribery, consumer interests, science and technology, competition, and taxation (OECD 2000). In spite of significant effort by some companies and much public debate about CSR, there is limited evidence that these new guidelines have in any way supplanted profits as the dominant priority in corporate decision making.

In effect, what has evolved is a set of nested processes to assist with making appropriate decisions related to new technologies. One can see them only by peeling back the layers of control, much as one might peel the layers of an onion. It is relatively easy to see the role of individuals, firms, and government: individuals and families are at least partly driven by concerns about individual utility and welfare; firms are driven by profit; and governments engage in regulating the excesses and failures of markets, to avoid the perverse outcomes of unfettered individual action. However, a new set of less visible actors are injecting new values and processes into the system. In the first instance, there are 6415 international governmental organizations seeking to govern aspects of the market and society. The London School of Economics Group on Global Civil Society estimates that there also were more than 59,000 international NGOs in 2003, each attempting to mobilize the voice of a group in society that shares a common value or interest. Standards-setting processes and bodies are also proliferating – there are currently standards agencies in almost every country, a set of international organizations such as the International Organization for Standardization, and more than 17,000 specific international standards, many of which attempt to balance economic with non-economic concerns. Digging deeper, one can see a more foundational set of governors, including the accounting profession and their rules of accounting and audit, the bond rating system, which assesses risk and return, and an array of formal and informal structures designed by citizens, consumers, firms, and governments to elicit the opinions and views of society (Phillips 2007).

While it is difficult to disentangle the processes related to animal biotechnology from this larger effort to make commerce and markets responsive to both market and non-market signals, an examination of the context, theory, evidence, and opinions expressed by our focus group participants helps to unpack the conflicts and convergences.

The Existing Animal Industrial Supply System

The existing supply system for animals in developed countries has come a long way from the pastoral, hunter-gatherer, wild herd system that emerged in pre-history. The animals used for food or industrial processes today have all been selected or bred to meet human needs, either for specific output attributes or to fit into increasingly industrialized production and marketing systems. While it is tempting to simply extrapolate our experiences with genetically modified (GM) crops into animals, the nature of the product and the industry is qualitatively different.

Jared Diamond (1997) suggests that humans' first foray into organized animal husbandry might have been with snails, but quickly moved up the value chain, leading to domestication of dogs, goats, sheep, camels, horses, cows, fowl, pigs, fish, crustaceans, and a range of other animals with specific uses. As of 2006, the world had about 49 billion chickens, 1.35 billion pigs, 298 million beef cattle, 241 million dairy cattle, and 539 million sheep under management. About one-third of the chicken, pigs, cattle, and sheep are raised in developed countries. On an annual basis, approximately 260 million tonnes of meat are produced, about 41 per cent in OECD countries, another 40 per cent in Brazil, Russia, India, and China (the BRIC countries), and the remaining 19 per cent in lesser developed countries (LDCs) (FAOSTAT 2008). The share of production has changed significantly in the past twenty-five years. While total production in developed nations has risen 45 per cent and production in LDCs is up 20 per cent, the BRIC nations have raised production by more than 470 per cent. As a result, the BRIC share of total production rose from 16 per cent to 40 per cent in the past generation, while the share produced in LDCs fell from almost 30 per cent to under 20 per cent.

This meat is then distributed across the global market through extensive trade systems. In 2005, world exports of live and processed cattle, chickens, and pigs totalled almost US$34 billion, with an average of more than 80 per cent of that value being exported from OECD countries. Canada's 5 per cent share of global trade, dominated by sales of cattle and pigs, was somewhat depressed from the long-term average due to lingering difficulties in exporting live cattle to the United States as a result of bovine spongiform encephalopathy (BSE) in the Canadian herd.

Consumption patterns vary widely, largely based on the purchasing power of residents in various countries. Animal protein is a premium or preferred food in most parts of the world. Because it takes anywhere from 1.5 to 7 tonnes of cereals to produce one tonne of fish, chicken, pig meat, or beef, animal proteins are more expensive than cereals and tend not to be consumed as much by people with subsistence or low incomes. Consumption is generally low and sporadic in populations with per capita incomes below US$5000. Once a nation's per capita incomes rise above that level, there tends to be a large and sustained rise in demand per capita. China's recent surge in production and consumption of meat illustrates this point: as China's economy grew strongly in the past few decades, its production of meats rose more than 500 per cent, making China the world's largest producer and consumer of meat products in 2006. One reason animal products are in such demand is that they are a rich source of protein – meat, milk and milk products, eggs, poultry, and fish contain balanced levels of amino acids, and hence are called complete proteins. In contrast, most plant protein is incomplete, lacking one of a variety of amino acids, and diets based primarily on plant protein often lack micronutrients that are essential for body development and long-term health. The Food and Agriculture Organization of the United Nations (FAO) estimates that, in 2001 to 2003, the average citizen in the developed world consumed 3300 calories a day, with about 21 per cent derived from animal products and another 17 per cent from various fats and oils. In contrast, the average person in a lesser developed country consumed about 2650 calories a day, with only 12 per cent from animal products and 10 per cent from other oils and fats (FAO 2008).

The underlying structure of the animals system has become significantly industrialized in the past one hundred years. Animal husbandry initially was primarily directed and managed by farmers in highly localized systems. Now most of the industrial systems involve highly organized and structured relationships throughout the supply chain. Most of the animals raised for the commercial food system (except horses) are artificially inseminated, using selected genetics to help standardize the quality and performance of the progeny. The quality and distribution of semen is often closely held by private companies or nationally based producer groups, while the quality and safety aspects of the system are governed by standards agencies, brand managers, and regional or national regulators (Spriggs and Isaac 2001). The other inputs to on-farm production, including commercial feeds, pharmaceuticals, machinery and equipment, and other consumables, are often tightly controlled and managed by private firms. On-farm production historically was relatively idiosyncratic, but the application of computer technologies, modern management systems, and private capital has tended to intensify and standardize farm production practices. There has been a significant rise in the number of farmers producing animals or animal products under production or marketing contracts: production contracts define the process for producing the animal or its product, while marketing contracts define the nature of the marketing relationship. The United States has perhaps the most extensive use of contracts. In 2005, approximately 11 per cent of all US farms, involving approximately 41 per cent of total production, had some form of contractual arrangement (MacDonald and Korb 2008). While use of contracts is lower in Canada, it is still significant and rising.

Moving downstream, there is extensive horizontal and vertical coordination between farmers, feedlots, processors, branded product managers, and wholesale and retail operations, which varies depending on the species. The beef industry has shifted towards a small number of very large specialized feedlots, which are vertically integrated with the cow-calf and processing sectors to produce quality-assured beef. In 2003, it was estimated that the market share of the four largest firms (called the four-firm concentration ratio) in the packing industry was 84 per cent for steer and heifer slaughter (Harkin 2004). Hogs are similarly concentrated. As of 2004, the packing company Smithfield Foods, the world's largest producer of pork products, owned 808,000 breeding sows, accounting for almost 14 per cent of the US total. This concentration continues in the slaughter business, where the four-firm concentration ratio was 64 per cent in 2003. Four of the five largest pork producers in the country, owning nearly 1.3 million breeding sows in

2002, also operate hog slaughter facilities and/or feed mills, in some cases controlling the product from the birth of the pig to delivery of pork to the freezer case at the local grocery store. The industrialization of poultry is almost complete. Ninety-five per cent of US poultry is produced under vertically integrated conditions, entirely in the hands of less than forty firms. The four-firm concentration ratio for broiler slaughter was 56 per cent in 2003. At the end of the supply chain, countervailing concentration has emerged. The five-firm concentration of supermarkets in the United States in 2004 was 46 per cent nationally and even higher in regional markets. Four-firm concentration for supermarkets in the four largest US metropolitan areas averaged 73 per cent in 1996.

Globally, food production and marketing has become more integrated. The sale of packaged food products generated total revenues of US$2.6 trillion in 2005, 39 per cent in Europe, 21 per cent in the United States, 31 per cent in the Asia-Pacific region, and 9 per cent in the rest of the world (Datamonitor 2006). Meanwhile, an increasing share of food is being purchased from large grocery stores rather than through independents or markets. More than 85 per cent of food sales in the United States and northern European countries, 65 per cent of sales in mid-income countries (e.g., Chile, Korea, Taiwan, Mexico, the Philippines, and Thailand), and 40 per cent of sales in rapidly developing lower-income countries (e.g., Indonesia, China, Vietnam, India) are going through supermarkets, and the trend is increasing (Hughes 2007). Globally, eleven retailers – all from OECD countries – are uniting the world at the cashier. Three global retailers – Carrefour (France), WalMart (USA), and Tesco (UK) – each have sales in excess of US$100 billion from operations in more than two thousand stores in up to thirty countries. The other eight international retailers from the United States, Europe, and Japan each convey more than US$20 billion per year in multiple markets.

Against that backdrop of a concentrated and integrated supply chain, one can also see an array of technological and managerial changes that have been intensifying the industrial nature of animal production. An array of innovations have already been adopted, including many non-transgenic innovations and a selection of biotechnology innovations that do not directly change the genome of animals. These technologies, processes, and institutions are being used to increase productivity or enhance quality (e.g., conventional selective and marker-assisted breeding, conventional veterinary drugs and new biologics, intensive livestock management systems, growth hormones, nutritionally enhanced feeds, and testing kits), to reduce the environmental impact of livestock production (e.g., centralized feed and waste management in intensive livestock operations and use of transgenic animal feeds with lower phytate), to produce industrial or pharmaceutical proteins (e.g., Premarinand insulin), or for aesthetic purposes (e.g., household pets and race horses). Each of these innovations, and many more besides, has worked to tighten the supply-chain linkages and increase profitability and productivity.

Some might argue that this cascading set of innovations places farmers firmly on the 'technology treadmill,' with new technology generating more productivity and lower prices, creating the need for further invention and innovation. While this is threatening to many, it is not all bad. One respondent in the Agricultural Producers focus group noted:

> When Dad started in the poultry business in the '50s, it took 12 weeks to grow a bird that was three and half pounds. Today we do that in 35 days. And in the . . . 30 years I've been farming, we've . . . taken seven days out of that . . . You've gone from the 2:1 feed conversion to the 1.75:1 . . . Those are huge gains. Huge. (Agricultural Producer V4)

In response, another participant asserted:

> It's progress. To me it's . . . being able to produce a lot more food with a lot less resources over that period of time, which is progress for Canadians, it's progress for the world, it's *progress for his family*. (Agricultural Producer V5)

This 'progress' has come with attendant costs and risks. Many farmers have had to retool and find alternative employment, families have relocated from farmsteads and rural communities to urban centres, and intensive production has reduced land-use diversity and tended to concentrate effluent and pollution in selected areas. Meanwhile, international market integration has both stabilized prices and incomes and exposed farmers to new risks of market disruption. The recent outbreaks of BSE and foot-and-mouth disease in the United Kingdom, BSE in Canada, avian flu in Asia, and the related trade disruptions illustrate the effect of this greater integration.

The long-standing and apparently inexorable industrialization of the global food supply chain, combined with the rapid and accelerating application of new knowledge, new technology, and new capital into the animal production system, has raised a host of concerns about the status of animals, the impact of animal production on the environment and society, and the distribution of the benefits and costs of the commercial system. While these changes have contributed to the rapid expansion of the global food supply in the past generation, they have also created winners and losers, haves and have-nots, have antagonized or alienated groups of consumers and citizens, and have mobilized an array of values-based, faith-based, and interest groups in a debate about the future of animals in our society. In a very significant way, these changes highlight the divide in the agricultural sector between the 'industrial' part of the agri-food system and those who identify as part of the agrarian movement, as discussed by Paul Thompson in this volume.

The Economic Consequences of Transgenic Animal Production

There is a significant amount of research being conducted in the area of animal biotechnology. Some of these innovations may fit comfortably within the existing production systems, supply chains, and consumer preferences; others will be difficult to reconcile with expections about our food system. How well they fit will to some extent determine if they are likely to emerge, and if they do, what impact they might have on the economic and commercial landscape for animal production. On balance, participants from the Agricultural Producers focus group tended to view biotechnology in animals as a way to

> expand tools available to address balance of production as well as needs of the animal, and the consumer, our market demands . . . There's always a bit of a tug between whether you're doing something strictly for benefits of production, agronomic purposes, or whether it's for just addressing the market demands, whatever that might be. And then of course on the animal welfare side, whether there are things that have negative implications for the animals. And so I'm looking at the balance of those three things. And when you have a new tool, rather than get carried away with the tool itself, see if you can strike a balance between those. (Agricultural Producer V3)

Where new technologies or products do not meet obvious needs, they tend to be less acceptable.

Different applications raise different responses from within the agricultral industry, from the industrial actors in the supply chain, and among both conventional and alternative agricultural operators. In some cases, the responses are a result of the 'fit' of the technology into the economic and social system, while at other times, the technologies generate a wider array of social responses.

A number of applications are simply designed to increase productivity. Already there is effort under way to clone bulls, to modify animals to resist viruses and bacteria that impede productivity (e.g., brucellosis and SARS), to create animals that can convert feed more effectively to meat (e.g., salmon), to modify animals so that they produce more proteins (e.g., cows bred to increase protein in their milk or, with kappa-casein and beta-casein genes, to facilitate cheese making), and to transgenically modify the digestive structure of animals to reduce waste (e.g., the Enviropig, to reduce phytate pollution).

Most producers and industrial actors see this as a valuable and important reason for innovation. Participants from the Agricultural Producers focus group were adamant that this was appropriate: 'I'm a believer in high-tech agriculture and the necessity of continuing down that path' (Agricultural Producer V4). Participants from this group appeared more than comfortable with anything that improved 'production efficiency' such as 'increased rates of gain, of selective muscle masses that are more useful' (V3) and 'genetic [*sic*] superior animals . . . that will give you longevity and the traits that are much needed for the consumer' (V5). Ultimately,

> it's about more efficient use of whatever we feed our critters in the production of protein for the world . . . Along the way, I'm thinking of things like improved economic viability for industries, profitability at the farm level, including things like alternate revenue streams but also including the environmental sensitivity and being able to maybe produce 10 per cent or 20 per cent more pigs with, you know, 15 per cent less water and less manure and that sort of thing . . . We need the profitability and we need the economic side, but it's really about playing in the world protein market. (Agricultural Producer V2)

Some of these applications garnered support from participants in the Regulators focus group. The Enviropig, for example, was 'consistent with the historical uses of pigs in agriculture' (Regulator V6). In contrast, participants from the Regulators, Animal Justice, and Alternative

Agriculture focus groups expressed less enthusiasm and greater reservations about some of these applications. Participants from the Regulators focus group had concerns that these applications raised new or higher risks than alternative production options. Participants from the Animal Justice focus group argued that the applications did not do anything to handle the fundamental concerns related to the status of animals. Lastly, participants from the Alternative Agriculture focus group believed the applications did not represent appropriate technological developments for their production systems.

Some proposed or attempted modifications are intended to improve human health through safer or more nutritious food. GM animals have been created to resist diseases that can affect humans (e.g., BSE), while others have been modified to improve the nutritional quality of the resulting meats or products (e.g., omega-3 pigs and cattle with the spinach gene to lower saturated fats). In addition, animals have been genetically modified to act as bioreactors for pharmaceuticals (e.g., the insulin-producing cow) and produce industrial proteins (e.g., spider silk in goats).

These applications raised a greater array of concerns. While many of them generated interest in the Agricultural Producers focus group because they would create 'a new income stream through a new market' (V1), there was concern that consumer acceptance must be considered. One participant from the Agricultural Producers focus group noted that consumer acceptance must be 'very much taken into account by industry in making sure that there is acceptance, before moving ahead with the technology so as to make sure that we don't harm our markets, both domestically and internationally' (Agricultural Producer V1). Another participant noted that the underlying fear is 'the backlash could be potentially worse than anything . . . I mean we're an exporting country. It could easily knock out our exports' (Agricultural Producer V4).

The Economics of Innovation

New technologies have fundamentally altered the innovation process itself. In the agri-food sector, the innovation process has historically been quite linear. It often started with curiosity-based research in universities and government laboratories and was moved through the development, production, and marketing systems guided by a mix of public incentive and private profit. As a result, the research process historically was relatively narrowly defined and self-contained. The advent of biotechnology in the agri-food sector did three things. First, it created the potential to more finely target the research to specific market needs. Second, it made the research process far more complex, with no one individual or small group of individuals able to undertake the entire research process. The resulting innovation process has become more networked and fluid than previously. Third, combined with modifications in the patent systems around the world, it led to a privatization of genetics that has the potential to create both winners and losers.

This new model of innovation has led to much greater specialization in the research business, a rise in the importance of networking and alliances, and a greater concentration of research activity, ownership, and control. This process of intensification and specialization has not yet run its course in the animals sector. Animal biotechnology is currently less concentrated and controlled than plant biotechnology. King and Schimmelpfennig (2005), using US Patent and Trademark Office data from 1970 to 2000, showed that the big six multinational biotechnology companies (Dow, Dupont, Monsanto, BASF, Bayer, and Syngenta and their subsidiaries) dominated the plant biotechnology business – controlling 97 per cent of the plant technology patents and 68 per cent of the patents related to metabolic pathways in plants. A number of crossover technologies – metabolic pathways and biological processes, protection, nutrition, biological controls, and genetic transformation – are also disproportionately controlled by the big six. In contrast, the big six were minority owners of patented non-plant organisms, metabolic pathways and biological processes in animals, and pharmaceuticals. One can assume that as the animals-directed technologies move closer to market adoption, private investment will rise, to more closely mirror the concentrated ownership patterns in the crops biotechnology industry.

This issue of control is front and centre in the minds of agricultural producers. One participant from the Agricultural Producers focus group summarized this nicely:

> A risk I see, and a concern, is attempts to tightly control genetic resources through patenting in particular, in place of allowing a broad exchange of genetics and broad ownership by the breeders . . . And so you risk narrowing the gene pool very rapidly, effectively, and you could put the genetic resource in some harm. And you're relying on only a small group

of people then to effectively maintain the genetic health of the gene pool. (Agricultural Producer V3)

One Alternative Agriculture focus group participant was perhaps even more concerned, noting, 'If we continue as we likely will, on this trend of reduced diversity among the animals, then there's a loss – there's loss to organic farmers or regenerative farmers because they no longer have the same pool to draw from' (Alternative Agriculture stakeholder V3).

The Location of Economic Benefits and Costs

Generally, the value being generated by these technologies is shared between the supply chain and end consumers. Currently, and for the foreseeable future, most of the technological innovations are likely to be developed, adapted, and adopted for the animals production systems in OECD countries. The combination of relatively weak R and D systems in LDCs, ineffective regulatory systems, and incomplete supply-chain development will for the foreseeable future make these new technologies less valuable to producers outside the OECD (Serageldin and Persley 2000).

The introduction of biotechnology would contribute to the major industrial restructuring that has been transforming the animals sector in recent decades. The first-generation, productivity-enhancing innovations are likely to be integrated into current supply chains. Future generations of the technology that add value for the end consumer are more likely to cause some restructuring within supply chains, either causing changes to existing chains or allowing new supply structures to emerge. The opportunity presented by biotechnology to manage the research process to deliver custom products has presented an investment opportunity for some private companies while creating a threat or risk to many existing businesses. The difficulties in protecting and transferring technologies among partners, combined with the imperative to quality-assure the product through the production and marketing chain, also push firms and the industry to vertically integrate. Moreover, firms need to be able to access, translate, adapt, and adopt new knowledge and technology to get products through the regulatory system. All these factors tend to concentrate activity in a limited number of sites. As a result, the production system from basic genetics right through to the consumer is increasingly tightly managed and may become more tightly controlled by a small group of industrial networks located in OECD countries.

This type of system has the potential to create significant overall returns for both industry and society. Economic studies done over the years show that research in agriculture has yielded relatively high private returns and even higher public returns (Alston et al. 1998), but that farmers tend to get a smaller share of the returns on innovations that improve yield rather than quality, and their share is depressed when the related supply chain is imperfectly competitive (Paul Thompson discusses alternative conceptions of benefits and costs in the following chapter). When these findings are considered in the context of biotechnology-based products, five conclusions can be drawn.

First, while a wide variety of economic studies of public research in the agri-food sector have estimated the internal rate of return ranges between 20 and 95 per cent (Alston et al. 1998), it is likely that the returns are lower for biotechnology-based developments. While some selected projects to breed high-value attributes into animals may still have high returns, in aggregate the action of competitive private research 'races,' often subsidized by public programs, is likely to lead to excessive investment, accelerated creative destruction, and relatively low social returns (e.g., Malla, Gray, and Phillips 2004).

Second, past studies have shown that the gains from yield-enhancing innovations are often bid away by competitive farmers, thereby translating into lower prices. Hence, yield-enhancing biotechnology products are not expected to benefit farmers significantly over the long term. Nevertheless, early adopters will often gain a 'first mover' benefit, while some individual farmers may gain as their specific agronomic and management circumstances allow them to profit from the new technologies (Phillips and Khachatourians 2001).

Third, consumers tend to gain at least some of the benefit of new technology. Yield-enhancing technologies tend to lower prices of food, an unambiguous gain to consumers, while quality-enhancing technologies tend to involve attributes that consumers value more highly, so that their satisfaction rises. This is not to say that some consumers may not be disadvantaged. When buyers are not offered effective choice in the market, consumers who perceive negative aspects of a new product may be less satisfied after its introduction. Transgenic modification has tended to be viewed unfavourably, all other things being equal, and so GM animals could disadvantage some consumers. Similarly, quality-enhanced products sometimes push older, lower-quality products out of the market, leaving only higher-priced, higher-quality offerings for consumers. Individuals who either do not benefit from the new quality attribute or who are constrained by limited purchasing power could be worse off.

Fourth, studies suggest that market concentration in the input and output sectors can lead to inequitable distribution of any net gains. Given the high degree of corporate control in and around the animals sector, one would anticipate that a substantial share of any production benefits will be captured by those firms that hold market power, at the expense of smaller and alternative farmers and less well-placed industrial firms.

Fifth, a variety of studies conclude that quality-enhancing innovations benefit farmers relatively more than yield improvements because they enable producers to segment the market and increase demand for their product, thereby offsetting the price-dampening effects of greater production. Quality-enhanced and industrial-based animal products could therefore yield a higher return, some of which could accrue to farmers because they will need to be paid to produce and to market in the way that protects the quality of the product.

Ultimately, while one can usually safely conclude that technological change introduced into competitive markets generates benefits that are shared by producers and consumers, blanket conclusions are not possible. Imperfectly competitive supply chains, closely held intellectual property rights, and lack of informed consumer choice could either limit the overall gains or dramatically change the distribution of those gains.

The issue becomes more complex when one considers that most agri-food products are traded internationally. The share of animals and meat that is traded is likely to increase because the new technologies are changing national and regional comparative advantages. While agri-food production has been relatively capital intensive for almost a generation, the sharp rise in private investment in the sector since 1990 has now also made it relatively research intensive. Economic theory suggests that production will tend to locate in regions of the world where the local supplies of land, labour, and capital favourably match with the requirements of industry. Although in theory technology should flow as easily as products, in practice technologies do not disperse rapidly or completely. An increasing technology or productivity gap between low-technology countries and research-intensive countries is developing (Sachs 1998). In addition, trade barriers that impede the redistribution of agri-food production to the most efficient sites are being dismantled through international agreement (e.g., the World Trade Organization [WTO] Agreement in 1995); that will allow agricultural production and trade to be determined more by economic competition. As a result, production of research-intensive products is likely to concentrate in and around those research centres that provide economies of scale and scope, which currently are mostly in developed countries. At the same time, many biotechnology innovations in the meat sector are likely to exhibit 'product' attributes. These products in many cases will be produced only in one country (near the research centre and in countries with functioning regulatory systems and supply chains) and then traded globally. All these trends in economies of scale and scope, product attributes, and consumer demand increase the dependence of the industry on international trade (Hillyer 1999).

As a result, one could hypothesize that the majority of the production gains will be captured by producers in those countries where the technology is developed and used. This could disenfranchise and disadvantage producers in other parts of the world (mostly in LDCs), as they will be competing with more productive or higher-quality producers. As a result, they could be losers. While most producers and firms are focused on their role in the supply chain, some producers worry about the lot of others. One participant in our Alternative Agriculture focus group expressed concern that

> the increased control by the agribusiness and pharmaceutical sectors of agriculture leads directly to a greater economic strain on small farmers and markets . . . not just here in Canada but . . . on the global south as well where agriculture is much less technological than it is here in Canada . . . So we have to start thinking what we are doing here and how that affects agriculture globally. (Alternative Agriculture stakeholder V1)

Nevertheless, provided the products of these different supply chains are internationally traded, one should expect consumers in developing countries to benefit relatively more from lower prices and better quality than consumers in OECD nations. In the first instance, there are overwhelmingly more consumers in developing countries – approximately five billion in 2008. Moreover, food accounts for a much larger share of average family expenditures in LDCs, and meat products are often prohibitively expensive. Thus, any technology that increases supply or improves quality is likely to benefit those consumers relatively more.

One issue that often is not calibrated in economic discussions about new technology is the nature of externalities. Animal production, in particular, generates significant waste and pollution. Any technological change that can either increase yield without a corresponding rise in waste or can act to reduce that waste directly could have a significant impact on the cost-benefit calculations for a technology. The dairy hormone rbST, for example, raises milk output and feed usage without a corresponding increase in animal waste; combined with other production efficiencies, this has helped to slow and, some argue, reduce the pollution per unit of output in the US dairy industry (Ott and Rendleman 2000). Meanwhile, there is a concerted effort to reduce pollution in the hog business. On average each of the 100 million pigs slaughtered annually in the United States produces approximately 17.5 pounds of waste per day (Petkewich 2001). The Enviropig has been genetically modified to produce manure with 75 per cent less phytase – it is the abundance of phytase that can contaminate water supplies, causing algal blooms that decrease oxygen in the water and kill fish. These benefits will accrue wherever these animals are produced, provided it does not simply lead to greater geographic concentration of hog farming, which could worsen the environmental impacts of the activity.

While economic theory and studies of past technological changes give us signposts for the scale and distribution of the benefits and costs of new biotechnology animals, the outcomes are not certain. Boehlje (2006) has noted that producers and their lenders invest in production capacity only if there is assured access to processing plants that can pay competitively for products. This interdependence results in the development of production-processing centres and supporting infrastructure. The geographic location of such operations is currently influenced by economies of scale and scope and the logistics of bringing feedstuffs to livestock and shipping livestock products to retailers. But given that capital and technology are increasingly mobile and that trade in live animals and meats has been significantly liberalized in recent years, it is likely that global livestock firms will locate production-processing capacity in those countries with the optimal mix of economic factors. Over time, that may offer real opportunities to many of the better-governed developing nations, such as the BRIC countries, while production in parts of North America and Europe may lose out.

Regulation of Production and Marketing of Transgenic Animals

There has been significant debate about the rules and structures that should govern the production and marketing of transgenic organisms. Currently, the rules are a hybrid of public, private, and collective processes. This area generates significant debate, as the normative assumptions underpinning regulatory structures and decisions are often unexplored (Brunk, Haworth, and Lee 1991; Mills 2002; Phillips 2007).

The state defines the *de minimus* legal minimum standards for the introduction and use of new organisms into the food system and the environment. North America and Europe tend to represent the two extremes: North America pursues a lassiez-faire market model, and Europe has developed a social market model. Other countries span the spectrum between the two (Phillips and Khachatourians 2001).

It has been argued that traditional government regulatory intervention in North America is based on a legalistic approach, while in Europe it is focused on political control (Woolcock 1998). The North American legalistic approach prescribes government regulatory intervention only in reaction to market failure; any regulatory intervention must preserve the fundamental focus on market forces. Essentially, this means that regulations are designed to ensure market 'efficiency or effectiveness' in correcting an apparent market failure (Majone 1990). The discretionary decision-making power wielded by regulators is kept in check through the transparency and openness of the decision-making process (Woolcock 1998). This approach both provides public scrutiny and limits the influence of populist politics on regulatory decisions. Indeed, with respect to modern biotechnology, one might conclude that North American regulations have been pro-competitive, relatively quickly considering, and for the most part positively deciding on, applications of new technology.

The European political control approach to government regulatory intervention is traditionally dominated by concerns over the democratic accountability of the discretionary decision-making power of regulators (Majone 1994). In the interests of achieving accountability, the objective of market efficiency tends to be subordinated. Regulatory decision making usually resides with elected public officials to ensure accountability. As a result, regulatory interventions can be subject to interests which dominate the concerns of elected officials. Thus, while market failure is often corrected with respect to 'social dimensions' (European Commission 1983), the regulatory evaluations tend to be more protracted and the decisions can at times lead to more restricted access to markets.

In practice, four fundamental principles differentiate the regulatory systems for biotechnology products. First, domestic regulations focus either on the products created through the use of biotechnology or on the technology used to make products. The challenge is that biotechnology as a production and processing method cuts horizontally across many areas, including agriculture, forestry, fisheries, pharmaceuticals, medicine, and the environment. These areas are traditionally regulated independently and pursuant to specific, often divergent, mandates. Second, some jurisdictions chose to use existing vertical regulatory structures, while others developed new horizontal systems. Choosing to regulate the *products* of biotechnology allows for preservation of the independent vertical regulatory structures. Choosing to regulate the technology, per se, requires regulatory intervention that cuts across the many applications and divergent mandates. Third, while most existing regulatory systems acknowledge the need for precaution, there are a range of divergent interpretations that can lead to somewhat different decisions based on the same theory and evidence. Fourth, some regulatory systems are relatively discrete, closed, and professionalized, while others are designed to be more open, pluralistic, and democratic.

Regulatory choices largely determine which products will be produced and how they can be marketed. The divergent regulatory systems have led to significantly different outcomes. The North American model has yielded relatively quick and liberal access to the market for new technologies in the agri-food sector, while the European Union model has had a mixed performance, at times approving applications relatively quickly and at other times delaying or rejecting applications. Even once a product is approved for use, the rules of marketing vary. A key variable that divides the global market is how products of biotechnology are labelled. In the North American system, any product that is judged as meeting the minimum standards for safety assessment is generally allowed to be marketed, the only general restriction being that any significant nutritional or compositional change or new health risk (such as allergens) must be positively labelled. If the resulting product, albeit the product of biotechnology, does not have any measurable metabolic difference or impact on health and safety, then no additional labelling is required. In some cases firms have proactively labelled the absence of genetically modified ingredients (e.g., GM-free or organic labelled foods), and in a few cases firms have proactively labelled the presence of GM traits that they think consumers might value (e.g., functional food properties). In the EU model, in contrast, firms are required to proactively label if transgenic technologies have been used in the production of a good, regardless of whether it has led to any detectable effect on the end product. Meat from transgenic animals would need to be labelled under this rule. While there was some discussion in the European Union about extending the rules to include any meat or product from animals fed transgenic feeds, these provisions were never enacted. Somewhat unexpectedly, in practice there appears to be a greater range of choice of food products in North America, where consumers can find in most food classes a range of products that are unlabelled (and presumed to be GM), proactively labelled as containing GM ingredients, or labelled as certified GM-free (sometimes via the organic standard) (Kalaitzandonakes 2003). In the European Union, most processors, wholesalers, and retailers appear to have re-engineered their supply chains to assure the products they market in the EU are GM-free, which allows them to avoid making any specific labelling claims.

As transgenic animals enter the food chain, these regulatory and marketing rules will have a differential effect. If the experience relating to GM grains and oilseeds is replicated, North America (and a range of large agri-food producers in the developing world) will adapt, adopt, and produce increasing quantities of transgenic animal products (likely at a marginally lower cost), while the European Union and a range of other nations will delay introduction of the technology. This divergence in technological diffusion is likely to lead to shifting market shares and increasing trade disputes as costs and prices diverge between the GM-producing and GM-free areas of the world.

Conventional and alternative agricultural producers and industrial firms in Canada are conflicted about how to deal with this regulatory conundrum. One prevailing view is captured by a participant in the Agricultural Producers focus group: 'Government has got a responsibility to protect human health and enviromental health. Leave the rest to us. Get out of our way' (Agricultural Producer V4). But in almost the next breath, producers will express concern about consumer backlash.

Conclusions

The pending introduction of GM animals highlights a set of conflicts that resonate throughout our agri-food system, and perhaps throughout our economic system. At one level, it is about winners and losers. In that context, it is important to understand that the new technology is only one influence on an already highly organized system that is in the middle of a fundamental restructuring that predates and transcends the technology. The new technology offers cross-cutting threats and opportunities. As a product of the scientific and industrial establishment, biotechnology presents an opportunity for increased concentration and corporate control over the food chain, which could accelerate and amplify the trend of greater specialization in production and international trade. While this offers the prospect of greater volumes of higher-quality food at lower prices, that process will inevitably create real, identifiable winners, a less visible group of disenfranchised losers, and greater potential for conflict over the distribution of those gains and losses.

At a deeper level, the pending advent of biotechnology in the animals system highlights the inherent conflict between the narrow morality of the marketplace and broader and deeper societal concerns. This conflict can be nicely summarized in a number of quotes from our focus groups. Ultimately, most of the concern is not about the technology, per se, but about who is pushing it, owning it, benefiting from it, and controlling it.

Despite the relatively positive views of agricultural producers about the structure and direction of the global agri-food system and the role of biotechnology in securing their position in that system, there is deep-seated fear that the operation of the system is changing farmers' status and view of their self-worth. One participant from the Agricultural Producers focus group exclaimed, 'We become serfs unto our own farm . . . a businessman, an entrepreneur. You're a cog in the wheel' (Agricultural Producer V4).

Meanwhile, a participant from the Alternative Agriculture focus group concluded:

> The first point that I wanted to make is that I think that looking . . . solely at animal biotechnology is overly reductionist . . . I think we need to step back because I don't have a concern per se with the science itself. I really believe that . . . humans should push these boundaries of scientific exploration, but I have more of a problem with . . . commercializing the results, and I think that's where things are beginning to go awry because that is where . . . something very natural called human greed comes into play and especially manifested through capitalism and our . . . economic system that I think has given a lot of power to corporations . . . When corporations use science to maximize short-term gain, they are not acting in the

best interest of society's longer-term interests and future generations; they are not acting in the best interest of the environment or biodiversity . . . They are not acting in the best interest of those who aren't part of their corporation . . ., the vast majority of people who don't necessarily . . . stand to benefit from their particular drive to maximize short-term gain . . . So my criticism is based in the critique of the . . . corporate-driven economic systems. (Alternative Agriculture stakeholder V3)

Finally, a participant from the Animal Justice focus group concluded:

I can't say everybody's evil, with malevolent intent, out there. I don't believe that. But I think people are willing to let themselves be persuaded or come up through a system where they have to believe certain things in order to succeed, or they have to do certain things in order to succeed. Animal research is a good one. And so probably yes, some people do think they're doing well, but I would not accept that the guiding lights of the corporation are in it for that reason. (Animal Justice stakeholder V2)

The process of making the choices about which technologies to develop and use will inevitably be grounded in some economic cost-benefit calculus tempered by a more humanist orientation mediated through reflexive government and corporate social responsibility. The outcome of that process is far from clear.

References

Alston, J., Mara, M., Pardey, P., and Wyatt, T. 1998. Research returns redux: A meta-analysis of the returns to agricultural R and D. Environment and Production Technology Division discussion paper no. 38. Washington: International Food Policy Research Institute.

Alston, J., Norton, G., and Pardey, P. 1995. *Science under scarcity: Principles and practice of agricultural research evaluation and priority setting*. Ithaca, NY: Cornell University Press.

Boehlje, M. 2006. Economics of animal agriculture production, processing and marketing. *Choices: The Magazine of Food Farm and Resource Issues*. www.choicesmagazine.or/2006–3/animal/2006–3-08.htm.

Brunk, C., Haworth, L., and Lee, B. 1991. *Value assumptions in risk assessment: A case study of the alachlor controversy*. Waterloo, ON: Wilfrid Laurier University Press.

Datamonitor. 2006. Global food, beverage and tobacco, industry profile reference code: 0199–2059, April. http://www.marketlineinfo.com/mline_pdf/industry_example.pdf.

Diamond, J. 1997. *Guns, germs, and steel: The fates of human societies*. New York: W.W. Norton and Co.

The Economist. 2004. Does it add value? 13 November: 81.

Food and Agricultre Organization. 2008. FAOSTAT. www.fao.org.

Gilpin, R. 2001. *Global political economy: Understanding the international economic order*. Princeton: Princeton University Press.

Harkin, T. 2004. *Economic concentration and structural change in the food and agriculture sector: Trends, consequences and policy options*. Prepared by the Democratic Staff of the Committee on Agriculture, Nutrition, and Forestry, United States Senate, 29 October. http://www.sraproject.org/wp-content/uploads/2007/12/harkinconcentrationwhitepaper.pdf.

Hillyer, G. 1999. Biotechnology offers U.S. farmers promises and problems. *AgBioForum* 2(2): 99–102.

Hughes, D. 2007. The global food market and developments in the modern food industry in India. Agri-business forum, 13th International Economic Forum of the Americas / Conference of Montreal, 20 June.

Kalaitzanondankes, N. 2003. Regulating biotechnology: GM food labels. Ed. A. Eaglesham, S. Ristow, and R. Hardy. NBAC Report 15, Biotechnology: Science and Society at a Crossroad, 125–40. Ithaca, NY: NABC.

King, J., and Schimmelpfennig, D. 2005. Mergers, acquisitions, and stocks of agricultural biotechnology intellectual property. *AgBioForum* 8(2, 3): 83–8.

Locke, J. 1690 (1937). An essay concerning the true original, extent and end of civil government. In *Treatise of civil government and a letter concerning toleration*, ed. C. Sherman. New York: D. Appleton-Century Co.

MacDonald, J., and Korb, P. 2008. Agricultural contracting update, 2005. Economic information bulletin no. EIB-35. http://www.ers.usda.gov/Publications/EIB35/EIB35d.pdf.

Majone, G. 1990. *Deregulation vs. reregulation: Regulatory reform in Europe and the United States*. London: Pinter.

– 1994. *Interdependence vs. accountability: Non-majoritarian institutions and democratic government in Europe*. EUI working paper SPS 94/3. San Domenico di Fiesole: European University Institute.

Malla, S., Gray, R., and Phillips, P. 2004. Gains to research in the presence of intellectual property rights and research subsidies. *Review of Agricultural Economics* 26(1): 63–81.

Mills, L. 2002. *Science and social context: The regulation of recombinant bovine growth hormone in North America*. Montreal: McGill-Queen's University Press.

OECD. 2000. *The OECD guidelines for multinational enterprises*. Revision 2000. Paris: OECD.

Ott, S., and Rendleman, M. 2000. Economic impacts associated with Bovine Somatotropin (BST) use based on survey of US dairy herds. *AgBioForum* 3(23): 173–80.

Petkewich, R. 2001. GM pigs produce less phosphorous waste. *Environmental Science & Technology Online*: Technology News, 19 November 2001. http://pubs.acs.org/subscribe/journals/esthag-w/2001/nov/tech/rp_pigs.html.

Phillips, P. 2007. *Governing transformative technological innovation: Who's in charge?* Oxford: Edward Elgar.

Phillips, P., and Khachatourians, G.G. 2001. *The biotechnology revolution in global agriculture: Invention, innovation and investment in the canola sector*. Oxford: CABI.

Roberts, M. 2000. US animal agriculture: Making the case for productivity. *AgBioForum* 3(2, 3): 120–6.

Sachs, J. 1998. Global capitalism: Making it work. *The Economist*, 12 September.

Serageldin, I., and Persley, G. 2000. *Promethean science: Agriculture, biotechnology, the environment and the poor*. CGIAR Secretariat, World Bank.

Smith, A. 1776. *An inquiry into the causes of the wealth of nations*. 5th ed. Ed. Edwin Cannan, 1904. London: Methuen and Co.

Spriggs, J., and Isaac, G. 2001. *Food safety and international competitiveness: The case of beef*. Wallingford, UK: CABI Publishing.

Woolcock, S. 1998. European and North American approaches to regulation: Continued divergence? In *Towards rival regionalism? US and EU regional regulatory regime building*, ed. J. van Scherpenberg and E. Thiel, 257–76. Baden-Baden: Nomos Verlagsgesellschaft.

5 The Farmers: The Agrarian Critique of Industrial Agriculture

PAUL B. THOMPSON

About twenty years ago, a small group of North American environmental and farm policy activists convened an informal collaborative to study the potential impact that then new methods of gene transfer would have on agriculture and to consider how activist organizations should respond. Known as the 'Biotechnology Working Group,' they included individuals such as Hope Shand, who would later be with The ETC Group, a Montreal-based organization that has been active in opposing crop-based biotechnology in the developing world, and Jaydee Hanson, who became the Center for Food Safety's leading spokesperson in opposition to animal cloning. Four participants produced a now obscure report entitled *Biotechnology's Bitter Harvest* that summarized key concerns and recommendations (Goldberg et al. 1990).

Although many of the specific environmental concerns noted in the report pertain to crop biotechnology rather than animals, the thinking in the Biotechnology Working Group was strongly influenced by the debate over the recombinant animal drug rBST, more popularly referred to as 'bovine growth hormone.' rBST had been the target of widely read research by Cornell University economist Robert Kalter (1985) showing that the drug would trigger a round of concentration (i.e., fewer and bigger farms) in the dairy industry. As such, one principal thrust of the Biotechnology Working Group's concern dealt with the likely impacts on small farmers and rural communities. Their report raised sharp questions about the influence that biotechnology would have on research at agricultural science institutions as well as on private companies that provide key farming inputs such as fertilizers and seeds. There was a fear that biotechnology would exacerbate trends associated with chemical pesticides with the advent of technologies that gave large-scale producers competitive advantages over smaller ones, and of production contracts that centralized control over farm decision making.

Despite the title's reference to a 'bitter harvest,' the 1990 report was not unilaterally opposed to the deployment of genetic engineering (and cloning of farm animals was not then on the horizon). These activists were calling for agricultural technologies focused on strategies identified in a 1989 report from the US National Research Council entitled *Alternative Agriculture* (NRC 1989). Techniques such as more complex crop rotations, composting, and pest control methods that would eventually come to be associated with organic farming were presented as alternatives to input-intensive farming methods, but the Biotechnology Working Group did not reject biotechnology as categorically incompatible with such alternative methods in 1990. Yet for the most part, specific questions raised by the Biotechnology Working Group went unanswered by mainstream agricultural science. Their request that university and government research labs offset research on biotechnology with research on alternative pathways was ignored. Thus rebuffed, these activists undertook a more strategic course of opposition to biotechnology in the 1990s, cultivating public opposition through a variety of campaigns that highlighted food safety uncertainties and that imputed a wide range of environmental risks to all types of genetically modified organisms (GMOs). Sociologists William Munro and Rachel Shurman (2008) attribute the eventual social movement in opposition to biotechnology to the leadership of individuals who were either in or influenced by the Biotechnology Working Group.

It might therefore be well for people schooled in the upshot of antibiotechnology campaigns carried out in the decade after GMOs began to appear in the food supply to understand the issues that motivated many of the activists who launched those campaigns. Doing so involves a foray into the contested terrain of agricultural ethics. In what follows, I first provide a framework for understanding the philosophical underpinnings of various claims and counterclaims about the ethical justifiability of present-day agricultural production methods. This framework is drawn from the writings of people arguing for or against industrial farming methods in various ways. However, several focus groups convened to reflect on animal biotechnologies can also be interpreted in light of this framework, as the second half of the chapter shows. The basics of agricultural ethics indeed matter for biotechnology, even if the implications are as mixed as the Biotechnology Working Group's report.

Whither Ethics and Agriculture?

Peter Phillips's chapter in this volume describes the 'technology treadmill' whereby many new agricultural technologies cause farms to become fewer and larger. In an important sense, the question for this chapter is, Why should we care? It is not simply a matter of one's personal feeling for the plight of small farmers or the preferences of activists that is at issue. When I first began teaching a course on agricultural ethics in 1982, a professor in agricultural economics accosted me one day for what he presumed was my naive and nostalgic advocacy of family farms in the course. In fact, I had not conceived of agricultural ethics as having anything to do with family farms, and had developed a course covering world hunger, animal welfare, and the environmental issues associated with agricultural chemicals. But my colleague's too strenuous protest tipped me off. The issue that agricultural economists refer to as 'farm size distribution' was and is central to agricultural ethics.

The percentage of farmers among the overall population in industrializing countries has been getting smaller for a long, long time. At the same time, the average size of farms has been getting larger, whether measured in acres or hectares, numbers of animals, or the value of the food and fibre commodities produced. These farms have also become much more specialized. Most large farms now produce just one or two things, whereas farms of yore would have produced a dozen or more crops or animals both for home consumption and for sale. There are exceptions to these trends. *Very* small farms have actually prospered throughout North America during recent times, though these operations seldom derive all of their income from growing crops and raising animals. Nevertheless, the trend to fewer and larger farms has been a hallmark of industrialization.

The reason my former colleague cautioned me against a naive endorsement of small farms is that in his view industrialization has forever changed the way we understand agriculture. He was implicitly endorsing what I have elsewhere called an *industrial philosophy of agriculture* (Thompson 2001). According to this view, whatever might have been the case in the past, we should now see agriculture as just another sector in the industrial economy. This means that society is best served when farmers, ranchers, and other animal producers make their products available at the lowest possible cost. They should not, of course, impose costs on others in order to do so; there are still important ethical issues that must be addressed in an industrial philosophy of agriculture, environmental impacts or fairness to workers and animals themselves among them. But we now live in an industrialized world, and perhaps we should simply face up to that fact.

I have called the opposing point of view an *agrarian philosophy of agriculture,* though others prefer terms such as alternative or multi-functional agriculture. This is a view that sees agriculture as performing a social function above and beyond its capacity to produce food and fibre goods such as meat, milk, leather, or wool. The exact nature of these broader social functions may be difficult to specify, and the reasons for regarding them as ethically significant may seem obscure at first. Yet it is very clear that some people – a minority, perhaps – have very strong feelings both *for* small-scale and traditional or diversified family-style farming and *against* the larger and more specialized farms that are typical of the industrial era. The strength of these feelings and the intensity with which they are expressed is alone a reason to see the opposition between industrial and agrarian views of agriculture as a debate with ethical and philosophical dimensions. It is this debate that becomes the heart of this chapter.

Industrial Agriculture: The Case for Specialization

Perhaps the trend towards fewer, larger, and more specialized farms is a good thing. This is the view of Jeffrey Sachs, for example. Sachs is the director of the Earth Institute at Columbia University and an economist known for his work on problems of poverty reduction, debt cancellation, and disease control for the developing world. Sachs describes the talents of the traditional diversified farmer as 'truly marvelous. [They] typically know how to build their own houses, grow and cook food, tend to animals, and make clothing.' But he goes on: 'They are also deeply inefficient. Adam Smith pointed out that specialization, where each of us learns just one of those skills, leads to a general improvement in everybody's well-being' (Sachs 2006: 37). Farmers who concentrate on raising just one or two types of crop, or who raise just one species of livestock, can become especially adept at the skills needed for that kind of work. They can exploit whatever advantages their soils or weather conditions might afford for a particular production activity to their fullest. This kind of specialization allows them to produce at the lowest possible cost.

The *ethical* argument for such specialization is that when goods are produced at the least possible cost, they can be sold at the lowest price. This means that people who eat (i.e., all of us) spend less on food, freeing up more of their income for other things. Low-cost food is especially important for the poor because it is those with the least to spend overall who can least afford to spend a larger share of their income on food. There are thus two ethical arguments at work here. First, selling *anything* for less is good on utilitarian grounds because when people have more discretion to spend elsewhere, they can allocate their spending to those things that they see as making the greatest contribution to their personal well-being. In this sense, technologies that improve efficiency serve 'the greatest good for the greatest number.' Second, selling *food* for less is good on egalitarian grounds, since it is an efficiency that is of relatively greater importance to the poor than to the rich. Technologies that reduce the cost of necessities (like food) are ethically better than technologies that reduce the cost of luxury goods (like entertainment) because they are especially beneficial to people with lower income. They tend to equalize life's chances, even if only in a relatively small way.

If the technology treadmill results in fewer, larger, and more specialized farms, it also results in a general lowering of prices for food consumers. Although the farmers who disappear from farming when the treadmill turns are the 'losers,' the winners are very numerous (all of us), with the comparatively disadvantaged among us winning comparatively more. So especially in circumstances where those displaced from farming can find employment in other fields, the technological treadmill looks like a bargain to development-oriented economists like Jeffrey Sachs. The economic details can complicate the arithmetic of this moral calculation substantially, as Peter Phillips's chapter illustrates. Benefits captured by biotechnology companies or the food industry may not be passed on to consumers. Details such as these matter a great deal for any final analysis of animal biotechnology, but the point here is to get clear on the underlying ethical logic of the case for specialization.

Notice that this argument does not treat agriculture differently from any other sector of the economy. Specialization in *any* sector can lower cost. Efficiencies in manufacturing can lower the cost of automobiles or video games. Efficiencies in energy production will be welcomed by all, and may be almost as valuable to the poor as efficiencies that reduce the cost of food. More efficient health care technology would, like cheaper food or energy, find moral support on both utilitarian and egalitarian grounds. Policies that prevented the introduction of more efficient technology in manufacturing, energy, or health care would be suspect. Have corporations, oil companies, or the health care lobby interfered in the political process to protect their profits? If this is how we would react to protecting the economic interests of those who produce health services, energy, or manufactured goods, why would we see protecting the interests of those who produce food any differently?

The argument for specialization applies across all sectors of an industrial economy. If we think that a norm of specialization and encouragement of the most productive technical means should not be applied in farming or livestock production, we are claiming that farmers are a special case. Some reasons for thinking they might be will be discussed below, but it is important to recognize that Sachs's adaptation of Adam Smith creates a burden of proof for those who think that agriculture is different. The view that agriculture is *not* different from every other sector of the industrial economy sees more productive technology as a good thing on both utilitarian and egalitarian grounds. This is the industrial philosophy of agriculture.

Of course it goes without saying that there is no true efficiency when there is no true reduction in cost. Lower prices can sometimes be achieved by deferring maintenance or investment costs. This means that these costs will have to be paid sometime in the future, and the lower price does not reflect a true increase in efficiency. Lower prices can also be achieved by forcing costs upon third parties. A company that lowers costs by lowering wages can sell for less, but it is simply the workers, rather than the company, who are bearing this cost. In these cases, lower prices might benefit consumers, but they do not reflect the true costs of production. Considering such cases is a critical part of agricultural ethics, and that is where the discussion must turn next.

Industrial Agriculture: The True Cost of Food

Some critics of industrial agriculture argue that the relatively low price consumers pay for food and fibre goods in industrial societies conceals a host of hidden costs. Jeffery Sachs praises Green Revolution technologies in agriculture for increasing the yields of farmers in India (Sachs 2006: 177), but Vandana Shiva (1991) argues that such calculations neglect the way that older farming methods returned nutrients and maintained soil structure. Soil-depleting practices defer costs into the future, especially if rising energy costs also increase the cost of artificial fertilizers. Bill McKibben (2007) picks up on the energy theme, noting that the alleged efficiencies in the industrial food system conceal the fact that enormous quantities of fossil fuel energy are used not only for synthetic fertilizers, pesticides, and farm machinery, but in transporting food commodities from distant markets. Our society's reliance on inexpensive fossil fuels has institutionalized a system that is imposing costs on future generations through practices that deplete resources and stimulate climate change.

These points are also noted by Michael Pollan (2008), who adds the more straightforward point that animal production in the United States and Canada is made to seem cheaper than it is by taxpayer-supported farm subsidies paid to grain farmers. These price support payments have lowered the costs of animal feeds far below the level that would have been established simply by the economic forces of supply and demand. Other costs take the form of direct harm to public health. Yale professor John Wargo (1996) has argued that weak laws regulating agricultural pesticides in the United States have unaccounted costs in the form of long-term health problems and the attendant medical expenditures allocated to addressing them. To the extent that the efficiencies extolled by Sachs depend on tax subsidies or entail costs in the form of harm to health, there are costs imposed upon present-day generations that have not been reflected in the prices paid by food consumers. They are, in the parlance of economics, 'externalities.'

There are also more philosophically controversial costs associated with industrial animal production. Modern feedlots and dairy operations utilize feeding practices that speed weight gain or increase milk production, but at the expense of gastric distress and increased rates of diseases such as mastitis. Modern egg production places hens in crowded conditions where they cannot express instincts to flap their wings, bathe in dust, or build a nest. Pregnant sows are kept in narrow crates throughout the gestation period where they are unable to even turn around. Broiler chickens have been bred to have such exaggerated breasts that they suffer from skeletal deformities. Gene Bauer, the founder of Farm Sanctuary, argues that these are costs imposed upon the animals themselves, and not accounted for in the prices that consumers pay for animal products (Bauer 2008). Animal biotechnologies have the potential to both mitigate and exacerbate some of these impacts on animals. In chapter 8 of this volume, Lyne Létourneau discusses the philosophical issues involved in taking animals to be moral subjects, but it is clear that to the extent we regard harsh impacts on animals as having moral significance, these, too, must be counted as among the true costs of the way we farm.

Now, the more precise way of understanding costs that are not reflected in the market price of a good is open to a number of different ethical interpretations. The language of cost may suggest a form of benefit-cost weighing that is familiar to utilitarian ethical thinking. Here, benefits from lower food cost might be thought to offset costs, and the crucial ethical questions revolve around the way that we estimate the value of costs and benefits, and then aggregate them in an attempt to determine whether or not the value of costs outweighs the value of benefits. A utilitarian might ask whether the benefits to consumers (especially poor consumers) exceed costs in taxes or medical expenditure, as well as costs for future generations or animals. Other social scientists attempting to take a more ethically neutral stance will say that the problem is one of deciding which trade-offs to make.

But one might also argue that certain types of harm should not be subjected to trade-off thinking. In the libertarian tradition of ethics, harms to health or liberty would be regarded as overriding benefit-cost calculations. A situation (such as appears to exist now) where affected parties lack legal rights to prevent such harms would be regarded as morally unacceptable. The libertarian approach to ethics is grounded in a view that regards social (or governmental) restrictions on individual action as acceptable only when they can be rationalized as ultimately necessary to protect others from harm. Citizens are justly compelled to respect each others' liberty and property on this view, but acts of beneficence are morally legitimate only on the condition that they are undertaken voluntarily. All the costs described above would (arguably) qualify as violations of libertarian moral rights, though there are difficult conceptual issues to work out with respect to animals or future generations.

Given the multiple ways in which we might approach these views, there is much more ethical work to do in specifying how to approach the unaccounted-for costs noted by Shiva, McKibben, Pollan, and Wargo, but these important issues will be set aside for now. A different philosophical point is worth noticing in this context: While this litany of externalized costs serves as an important qualification on the case for specialization, the moral force of these claims is wholly consistent with the industrial philosophy of agriculture. Recall that this philosophy states that farming should be treated just like every other sector of the economy. Certainly there is cause for concern about adverse impacts on health or unjustified tax subsidies in industries such as energy or manufacturing. (Indeed, these are frequently the subject of moral critiques of these industries.) Similarly, there is nothing particularly unique about agriculture when it comes to costs imposed on future generations. Costs to future generations in the form of climate change would appear to be due primarily to transportation and household energy consumption, for example.

Activities such as the use of animals in medical experiments have been the focus of protest by advocates for animal rights. Agriculture is a big user of animals, but the claim that we should account for all the costs of an activity, include those to non-humans, is consistent with the main thrust of industrial thinking. So all these critiques levelled at industrial-style farming are well within the purview of a philosophy which states that agriculture should be viewed as an ordinary sector of the industrial economy. Hence one might conclude that if there is a disagreement between someone like Sachs, who advocates more industrial farming, and critics who seem to want less of it, it is a disagreement about the outcomes that specialization has actually produced. It is, in other words, a dispute about the facts and not about the moral principles that should apply when thinking about agriculture.

The Agrarian Mind

In contrast to the industrial view, an agrarian philosophy takes agriculture to have moral significance that extends well beyond that of industries such as transportation, manufacturing, or even health care. For example, the ancient Greek poet Hesiod (ca. 700 BCE) saw farming as having a religious purpose. His Zeus was an immanent god, thoroughly integrated into nature and the source of its unity. The seasons, soil, and water are themselves divinities begotten by Zeus that establish a place for human beings. For Hesiod, only farmers dependent on seasons, soil, and water can hope to attain piety with respect to these divinities. Farming is the way that human beings justly occupy a place in the divine order, which is, in Hesiod's world, the *natural* order. And it is the gods' intention that this place be fraught with work, toil, and risk. The harms that befall humanity as a result of natural causes are but instances of divine justice. Warfare, violence, and trickery, in contrast, are unjust in Hesiod's poetry because they short-circuit the gods' intended route to material rewards. In Hesiod's world, these human artifices will eventually (and inevitably) be repaid with misery and loss. Agriculture is thus the singular practice by which humanity makes its way in the world in a pious and morally just manner. It is, of course, important to Hesiod that agriculture bring forth the food and fibre goods that are the focus of industrial thinking, but this is only the tip of the iceberg for his understanding of the moral significance of farming (Nelson 1998).

Hesiod's poetry provides one dramatic example of the way that one might understand agriculture to have moral purposes that extend well beyond the efficient production of commodities. Throughout history, agrarian philosophies have valorized an enormous variety of goods and ideals stemming from agricultural practice, and such ways of thinking were commonplace in North America until recently. In 1948 political scientist A. Whitney Griswold found it necessary to devote a book-length monograph to refuting the view that truly democratic societies were inherently agricultural in their economic organization and political structure. Griswold made Thomas Jefferson, third president of the United States and author of the US Declaration of Independence, into the symbolic progenitor of the myth (Wunderlich 2000), though it is arguable that other figures of the time were more committed to it. Benjamin Franklin, for example, argued that farming produced a moral personality more inclined towards honest dealing and loyalty to one's fellow citizens than did the city trades (Franklin himself was a printer and publisher). The upshot: only a society of farmers will develop the personal habits and virtues necessary for self-rule (Campbell 2000).

This kind of argument is still with us. Historian Victor Davis Hanson, reputed to be former US vice-president Dick Cheney's favourite author, has expressed similar views, though in a vein suggesting that the lack of contact present-day Americans have with farming has corrupted their political values. Hanson (1996) believes that only farmers have a true understanding of how secure property rights lend support to virtues of patriotism and citizenship. Authors on the left include Brian Donahue, who believes that reconnecting people to the land through gardening and eating patterns that stress both seasonality and local production is essential to the development of a moral personality that appreciates what it takes to support diverse but culturally integrated and unified communities. Donahue (2003) believes that 'urban agrarianism' is most critical for the type of political mentality that highly integrated and localized eating and farming practices create. A deep sense of mutual interdependence and common purpose emerges in localities where people interact in the production of food.

Poet, novelist, and essayist Wendell Berry is unquestionably the most articulate advocate of agrarian philosophy in our own time. Berry confronts the claims of specialization directly, arguing that the division of labour has created lives whereby people are unable to see the larger wholes in which both human relationships and exchanges with nature acquire their meaning. The fragmentation of contemporary life corresponds to a vision of human beings as 'choice-makers' who move from transaction to transaction, evaluating choice in atomistic terms, as if choices and the places in which people live and work did not form a larger, more integrated form. In place of this fragmented consumer lifestyle, Berry advocates a return, however partial it must be, to practices embedded in and emanating out of a commitment to a given place. For Berry, the communities that have come closest to achieving true community (and true stewardship of the environment) are traditional farming communities (Berry 1977; Smith 2003). Thus Berry advocates, if not a literal return to agrarian lifestyles, then at least the deliberate cultivation of an ethical mentality that locates our ideals of polity, community, and environmental responsibility in agrarian ideals.

Understanding Agrarian Philosophy

Notice how criticisms levelled against industrial agriculture take on a different meaning when framed within the context of the agrarian mentality. Advocacy of local food is a particularly salient example. In stressing the energy costs associated with the long-distance transport of foods and the role that this ecologically needless expenditure of carbon plays in global warming, the criticism forcefully suggests that a more comprehensive accounting of environmental costs associated with industrial farming would produce a different verdict in terms of ethics. While the low food prices of the industrial food system are ethically good things for all the reasons noted by Jeffrey Sachs, there is more to the story. When the long-term environmental costs associated with energy consumption are factored into the equation, the cost-benefit ratio appears as not so attractive. For those inclined to a libertarian way of thinking, it could be argued that these benefits to present-day consumers are being obtained by imposing costs and risks on future generations, generations that have necessarily been unable to give or withhold their consent to this 'bargain.' Either approach can generate an ethical critique of industrial food production.

But an agrarian is more concerned with the way that a local food system embeds people in practices where their commerce with nature and with each other will create an enduring sense of place. Even people who buy most of their food in farmers' markets or through cooperative arrangements will encounter the same people repeatedly, week after week. They will build bonds with them, and the need for honesty and mutual respect is critical in such repeated encounters. Furthermore, these are the people who will be growing the food. Consumers learn the rhythm of the seasons, and they will know what grows well under local conditions. They can inquire about the condition of the land and animals under the farmer's care. The agrarian hope is that these kinds of localized transactions will gradually develop into an affection for the people and the place in which one lives, and that this affection, this sympathy, will in turn mature through the constant repetition of these rhythms into full-fledged habits of character – virtues, if you will.

The overriding moral concern that emerges from the agrarian mindset is thus one focused on the way that quotidian material practices establish patterns of conduct that are conducive to the formation of certain habits. These habits become 'natural' to people who engage in them repeatedly and become the stuff of personal moral character. When such habits are shared throughout a locale, they form the basis for community bonds and become characteristic of people living within that locale. Food production and consumption has been one of the activities most strongly tied to repetitive material practice. Furthermore, these localized practices are shaped by tradition and geography, by soil, water, and climate conditions. It is thus not surprising that moral philosophies focused primarily on the emergence and stability of virtues, community, and moral character would converge with a mindset that takes agriculture to have special moral significance.

In point of fact, many of the criticisms noted above as focused on the 'true costs of food' can be reframed in agrarian terms. Concerns about the long-term fertility of soils and for the ecologically adaptive characteristics of plant and animal varieties can be understood as an expression of the stewardship or husbandry that characterizes a well-functioning agrarian economy. To fault industrial systems for paying too little heed to the human practices that safeguard fertility and genetic diversity, as Shiva does, can be understood as a claim focused not on the *impact* for future generations, but on the need to preserve habits or virtues dedicated to land stewardship and animal husbandry. Concerns about the distorting effects of subsidy can be reconfigured as complaints about the way that repetitive material practices (the purchase and consumption of food) have themselves become warped by a dysfunctional economic environment.

Indeed, warnings about the dire consequences that will visit those who stray from agrarian habits of character are wholly consistent with the agrarian mentality. Hesiod's *Works and Days* is full of warnings for fools who neglect their farms and engage in 'grabbing' or indolence. These warnings should not be interpreted as Hesiod encouraging his audience to better calculate the true costs of these vices. The entire concept of a rationally calculating economizing mindset was wholly foreign to his outlook. For Hesiod (and for many since), those who try to operate outside the place laid for humanity are simply fools. The bad consequences that befall them are marks of their foolishness, events that confirm the flaws of their character. Of course, even good farmers can have bad luck; Hesiod's poem is full of examples. But in the case of good farmers, bad consequences do nothing to controvert their basic righteousness. Good luck is not a *reward* bestowed on the righteous for Hesiod. Rather, the fact that piety can fall victim to bad luck shows that being motivated by expected outcomes has very little to do with morality for Hesiod, as well as for most of the ancient Greeks. Only a grabbing fool would try to outmanoeuvre the gods, and it is the fool's fixation with playing the odds that is the very mark of his foolishness as well as his impiety. However strange this kind of thinking may sound to contemporary ears, it is worth noticing how Hesiod can predict bad outcomes from bad behaviour without also suggesting that these bad outcomes are the *reason* for thinking that the behaviour is bad.

Thus, understanding agrarian morality requires an agrarian-based talk of risks as making a different point about bad outcomes than that associated with the risk-benefit trade-off thinking of the industrial mind. There is thus a complex moral dialectic at work in the agrarian critique of industrial agriculture. On one level there is the push for efficiencies and specialization associated with liberal ideals of social welfare and economic growth. This push is countered by a recitation of risks, externalities, and costs that have been neglected in the moral accounting that supports this push. But the moral ideal that undergirds industrial thinking fully accepts the need for a full accounting, even while recognizing that people will differ in how they evaluate the trade-offs among costs, risks, and benefits. There is thus a 'pro and con' debate over the current shape of agricultural production taking place *within* the industrial paradigm. In this debate partisans of mainstream agriculture argue with critics who recite a litany of unaccounted-for costs. At a totally different level, there is a philosophical challenge to a conception of ethics that is dominated by decision-focused trade-off thinking. This view sees ethics as more appropriately focused on the cultivation of habits, norms, and institutions that will obviate the need for constant weighing of costs and benefits.

Habits, norms, and institutions are the fabric of social life. They constitute the individual within a community. Agrarian philosophies see food and farming as especially critical to the formation and reproduction of community identity and solidarity in virtue of the ubiquity of food practices and their tight connection to nature. People who take this perspective might also list risks and unwanted outcomes, but, like Hesiod, they do not think that getting the trade-offs right is an adequate moral response to risk. Rather, the tendency to think of food production in terms of trade-offs is simply evidence that our whole cultural system – our philosophy of agriculture – is out of whack. It shows the need to reorient our food system towards practices that produce and reproduce the global moral orientation to nature that reminds us of our place in the natural order.

Industrial and Agrarian Themes in the Focus Groups

Participants in stakeholder focus groups conducted for this project echo many of the concerns noted above. 'Production efficiency' (V3), for example, was the first thing mentioned when participants in the Agricultural Producers group were asked about the potential benefits of animal biotechnology. 'Cheaper ways of producing' (V1), 'more efficient use of whatever we feed our critters in the production of protein for the world' (V2), and 'being able to maybe produce 10% or 20% more pigs with, you know, 15% less water and less manure' (V2) were also given as examples of benefits. Although the group expressed a wide variety of rationales for rating the benefits and risks of specific examples, increases in production efficiency was among those mentioned as a reason for rating applications positively.

Figure 5.1: A framework for agricultural ethics 101

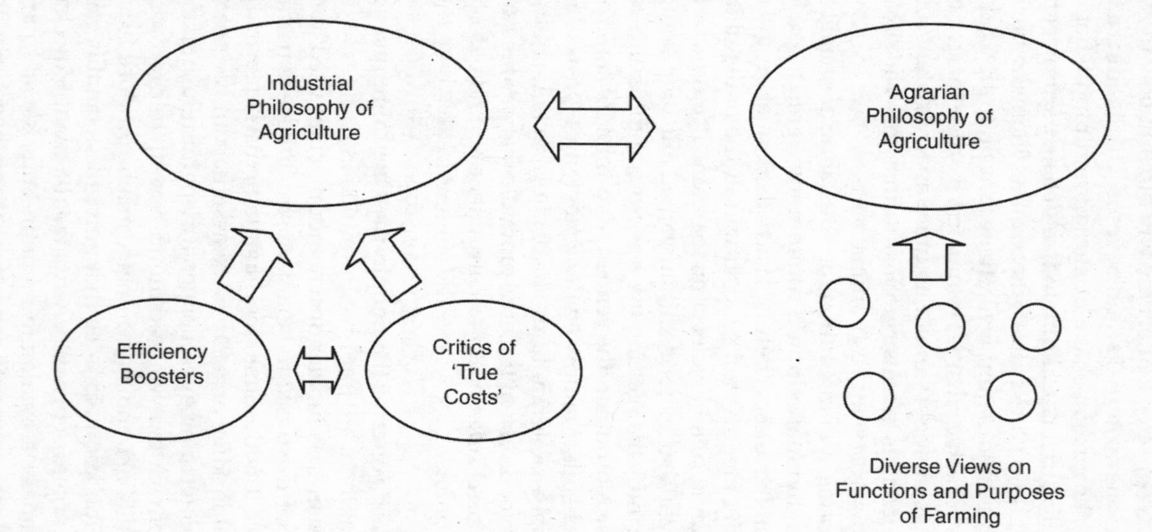

At the same time that participants from the Agricultural Producers focus group endorsed efficiency, they recognized that the benefits of increased efficiencies do not always accrue to producers. One participant expressed concern about 'corporate control' (V3), to which another responded, 'Another way of saying the same think is I think it's going to limit the ability of farmers to make money . . . Things that people sell to us are not based on [their] costs of production, they're based on potential benefit to us. And it actually narrows our profit opportunity as opposed to widen[ing] it . . . We become serfs unto our own farm, and [animal biotechnology] is just another step in that direction' (Agricultural Producer V4). This theme was endorsed by several in the group, including V3, who said, 'You are no longer a true owner' when technical changes shift decision making from farm to the technology providers. Thus, the Agricultural Producers focus group participants both exhibit an appreciation of the social benefit associated with increased cost efficiency, and also understand that the technology treadmill may prevent many of the benefits of more efficient production technology from being realized by farmers.

Participants in the Alternative Agriculture focus group also see increases in farm productivity as a putative benefit of animal biotechnology. One notes that the cloning of dairy cows 'with very productive genes' (V4) is a likely application of biotechnology, and another noted that using genetically engineered animals to lower the production cost for drugs would be a benefit to consumers. They are quick to add that the primary beneficiary will be the companies that own the technology, rather than animal producers or consumers, however. V4 also notes that 'the main hopes of the people who are doing the agriculture research is that there are improved production features, whether it is actually increased production or whether it is increased product quality. But I don't perceive that to be a very strong benefit particularly when you compare it with something like the human medical benefits.' Participants in the Alternative Agriculture group are, on the whole, rather negative about the prospects of genetic engineering and cloning for the food system. 'My only hope around biotechnology food products is that it doesn't happen. It's eliminated,' says one (V1).

As such, it is not surprising that participants in this group are more inclined to note risks than benefits from animal biotechnology. Risks to the welfare of animals are mentioned by several participants, including one who opines that biotechnology 'just facilitates the continuation of a cruel industry' (V1). This participant also notes a number of generalized concerns about the unproven nature of the technology, linking this to questions of food and environmental safety: 'There's also concerns, I think, around the contamination of the food supply.' Participants in this group conducted a lengthy discussion around the impacts on biodiversity. Participant V1 noted that centres of diversity are 'the most vulnerable from these kinds of technologies.' Another saw current trends 'sacrificing the wild salmon' (V3), and did not see the prospects of genetically engineered fish as an appropriate response.

Ethics, Regulation, and Animal Biotechnology

The expressions of concern in the Alternative Agriculture group might be said to 'hybridize' a number of concerns that could be expressed in a more 'purified' way. Industrial regulatory philosophy presupposes that the unwanted consequences associated with any technology (including a food technology) can be separated – purified – into distinct ethical categories like food safety, human health, environmental impact, and social consequences. Not only do distinct regulatory agencies of government deal with each distinct domain, distinct scientific disciplines are used in the risk assessments that inform decision making for each distinct regulatory agency (Thompson 1997). In this way, both information gathering and decision making become confined within disciplinary silos, a phenomenon that Norton refers to as 'towering' in reference to the discrete towers that separated different divisions in the old US Environmental Protection Agency (Norton 2005: 29–42). Those schooled in the disciplines of purification become exceedingly frustrated with ordinary people who run these distinct kinds of concerns together, seeing factors that might lead to concentration of agriculture as sources of uncertainty about environmental or food safety risks. They see this hybridization as a conflation of things that need to be kept logically distinct.

The very idea of risk can be seen to encourage such hybridizations, however, at least when one can free oneself from the industrial mentality long enough to think like an ordinary human being. In regulatory risk assessment, one maintains a narrow focus on specific biological hazards and the scientifically known causes that might allow such a hazard to materialize. In ordinary life, one can be placed at risk for reasons that have little to do with biology. Indeed, one of the most familiar ways to be 'at risk' is to be in a situation where other people can take advantage of you because they know something that you don't. Buying a used car became emblematic of a risky thing to do because sellers know much more about used vehicles than a potential buyer can ever ascertain. It is thus typical for an ordinary person to run all circumstances relating to vulnerabilities of any kind (including uncertainties and information inequalities) into a hybrid notion of risk. Vulnerabilities relating to trust and the uncertain or shady character of other people come sharply to the forefront in this way of thinking (Thompson 1997).

The agrarian emphasis on moral character and community solidarity presents an ethical framework aptly suited to the hybridization of potentially separable and distinguishable hazards, concerns, and outcomes. Expressions of risk that refuse to distinguish the discrete logic of objective hazards, felt vulnerabilities, and vague uncertainties may be articulated as distrust in the moral character or mentality of those to whom key decisions have been delegated. Agrarian ideals promote the vision of a well-integrated community in which people interact with one another over many years, if not generations. In such settings, people come to be known by their character, and the question of whether they can be trusted is never in doubt. Although agrarian societies may have their share of scoundrels, the intense sociability of agrarian reciprocity creates a situation in which virtues of honesty and reliability may be rewarded and reinforced more frequently than in those where commerce is routinely conducted amidst strangers. In agrarian societies 'risky' may be an adjective that describes people whose character does not inspire trust or confidence, and the goods or services offered by such persons become risky through a kind of moral contagion. Conversely, the recitation of hazards, vulnerability, or uncertainty associated with a given product or activity reveals a flaw in the character of those to whom responsibility for the product or activity is entrusted.

The agrarian moral framework is arguably rather unsuitable for governing technology in a modern industrial economy. Formal rules and regulatory procedures here have certainly replaced agrarian reciprocity in an instance of what Max Weber theorized as a process of rationalization. Ferdinand Tönnies describes this as a process in which *Gemeinschaft* (often translated as 'community') disappears and is replaced by *Gesellschaft*, or civil society. These themes in social theory tie the distinction between industrial and agrarian philosophy of agriculture to long-standing debates over the impersonality and alienation associated with modern societies (Thompson 2007). A widespread discontent associated with regulation may well be part of this. Thus, almost all the focus groups, with the notable exception of the Scientists and the Regulators, express some qualms about regulation, even while they recognize the need for formal regulatory processes in order to protect public health. It may be telling that only those who implement the logic of purification seem entirely happy with this approach to social relationships.

Ironically, the self-enforced neutrality of ethical perspective associated with respondents in the Regulators focus group emerges as seeming somewhat less neutral, given our portrayal of the issue here. Not only did several respondents in this group endorse the idea that the ultimate test for animal biotechnology should be in its net benefit to society as a whole, they also persistently suggested that agricultural and health applications should be viewed in similar terms. One stated:

> So one of my hopes is basically that once we've got these products and health and safety has been determined, at least in one jurisdiction, that it becomes accepted in others. Because there's a lot of – you know, my hope is that – the hope is that the benefits of these applications be widely recognized in the world and that we don't have – somehow we deal with, or these powerful ideological currents in the world that basically say government regulators and scientists can't be trusted and no application of biotech is ever going to be shown to be safe. (Regulator V3)

Although the quote is ambiguous, it might readily be interpreted to mean that health and food applications can be evaluated with the same overarching framework, and that the appropriate framework will give priority to scientifically determined measures of safety. This is *not* to say that people in this group were disposed to regard animal biotechnology favourably. Indeed, the Regulators focus group may have had more sceptical comments on the potential risks of animal biotechnology than any save the dominantly sceptical Alternative Agriculture and Animal Justice groups. Yet in being 'neutral' about the prospects of animal biotechnology, participants from the Regulators group evinced little sympathy for hybridization of risks or for the underlying logic of an agrarian philosophy of agriculture.

This still does not imply that one focus group can be thought of as representing an industrial philosophy of agriculture, while another is centred on agrarian ideals. It does suggest that the broad characterization implied that the framework developed here can be useful in articulating and understanding how concerns that move well beyond straightforward social consequences might become intertwined with worries about a technology's effect on patterns of ownership, economic decision making, and the concentration of economic power. It is possible to frame these issues *within* an industrial paradigm, and fairly straightforward ethical questions of social justice and public health are the result. It is equally possible to frame the issues that are bearing on personal character and community identity, and to see them as deeply (if vaguely) interconnected. Any reasonably complete ethical discussion of animal biotechnology will make room for both ways of framing the issues.

Social Consequences Redux

The participant in the Regulators focus group who hopes that 'once we've got these products and health and safety has been determined, at least in one jurisdiction, . . . it becomes accepted in others' (V2) may have been thinking about rBST, the animal drug mentioned at the outset of this chapter. rBST was found safe and effective and approved for use in the United States, but regulatory authorities in Canada and Europe reached a different conclusion. Although the putative basis for the different findings lay in different risk management decisions regarding animal health, many have speculated that the social impact that Robert Kalter predicted for dairy farms was the underlying reason for resistance to rBST among consumer groups and some farmers (Lacy et al. 1992). The issue here was the technology treadmill in the dairy industry – a technical change that hastens the transition towards fewer and larger farms. Kalter predicted another turn of the technology treadmill following the introduction of rBST, with further concentration in the dairy industry as the result.

As rBST entered the market in the United States after 1992, these predictions appeared to be being realized. Herd size is a reasonably good measure of scale in the dairy industry. Average herd sizes in the United States grew from approximately 200 to 2000 animals in the decade from 1985 to 1995, and by 2005, herd sizes in the range of 10,000 animals were not uncommon. But there was another important, if little recognized, confounding factor. The use of computers to automate and manage record keeping in dairy production made it comparatively easy to solve management challenges associated with an increase in the average herd size. Small farmers with herds of fewer than 100 cows could probably remember each cow. As numbers grew, it became important to use numbered tags and to keep written records of milk production and veterinary treatments for each animal. With computers to manage this task, the record could be generated simply by typing the number on an animal's ear tag into the database. Soon, barcode readers and RFID tags automated even this task. It is doubtful that any biological technology, including rBST, would have been deployed in the absence of computerized management abilities. It is likely (though no empirical research has been done to support this conjecture) that computing technology, rather than biotechnology, was the proximate cause of the size-distribution change that occurred in the US dairy industry from 1992 to the present.

Needless to say, consumers have not been up in arms protesting the use of information technology in the dairy industry, but what does this mean for the ethics of animal biotechnology? If this conjecture is correct, it suggests that mobilizing public opinion in support of agrarian values may well require something more than a straightforward technology-treadmill argument. It is reasonable to conjecture further that a combination of risk-based concerns about product safety, opposition from animal protection groups (though they, if anyone, should have objected to the computers), and the mixed 'yuk factor' and religious responses widely associated with genetic engineering and the cloning of farm animals is needed to mobilize public outrage. This, in turn, would suggest that the putative social movement concerned with a less industrialized food system or eating local foods may not, in fact, have a great deal of traction beyond a small minority of food consumers – consumers much like those represented in the Alternative Agriculture focus group. In other words, a substantial majority may well be quite satisfied with an industrial philosophy of agriculture. Agrarian philosophies of agriculture provide philosophically stimulating lines of analysis for animal biotechnology, but it is not at all clear that they really have much motivating force.

There are, however, two points to consider in reply. First, the influence of computer technology on the size and scale of dairying may be so well known among dairy specialists as to be a commonplace, but it is not widely known or well understood among the broader public. It is at least possible that anyone with agrarian leanings and a more sophisticated understanding of how information technology affects milk production would respond in a manner similar to the reception that greeted rBST in the 1990s, and that potentially awaits other animal biotechnologies today. Second, it is worth stressing that ethical disputes are not settled by opinion polls. The contest between industrial and agrarian philosophies of agriculture establishes opposing perspectives on the best way to think about the broader socio-economic impacts of agriculture and agricultural technology. The degree to which these perspectives actually motivate public opinion on animal biotechnology is not, in itself, evidence that weighs one way or another with respect to the ultimate disposition of the philosophical issue. Human slavery was widely supported in the late 1700s, but this does not incline us to think that it was acceptable from a *moral* perspective. We are more inclined to see slavery's acceptance as evidence of a widespread character flaw.

Animal Biotechnology and Moral Character

Many critics have suggested that the moral problem with animal biotechnology lies in the character of anyone who could think that doing something like that to animals was okay. 'Like that' is often left a little vague, but several authors have been especially articulate in explaining that it is not so much *what* is done to animals as it is the attitude with which it is done. This was the theme in Allen Verhey's important article 'Playing God and Invoking a Perspective' (1995), wherein religious language used to impugn genetic engineering is reinterpreted as language intended to suggest that pursuit of commercial gain and instrumental uses of life involve patterns of conduct that are intrinsically objectionable when it comes to one person's relationship to another, especially when that relationship is mediated by genetic technologies. Allan Holland (1995) made a similar argument about the acceptability of animal biotechnology, suggesting that the problem resides in the way that genetic manipulation leads us to think of animals as mere things, infinitely manipulable and at the disposal of human beings. The problem here is not a harm done *to* the animal, but the way people come to think of living beings as mere resources. Arguments of this sort that came up in focus groups are discussed in Harold Coward's chapter in this volume.

Autumn Fiester frames the case for a 'presumption of restraint' in genetic modification of animals by arguing that frivolous uses of genetic modification seeking entertainment, profit, or artistic expression illustrate a morally objectionable form of hubris, offend moral sensibilities, and risk harm to the animal subjects of biotechnical transformation. She makes extensive references to Alba, the genetically engineered green fluorescent art-project bunny, as well as to the GloFish. Clearly, the widespread dislike of such applications in focus group responses would lead one to think that Fiester has struck a nerve. Her emphasis on hubris and offence links her objections to the kind of character flaw arguments that come up both in conjunction with 'playing God' and in connection with agrarian ideals. Agrarians might agree that interactions between human and non-human animals are a potent ground for reciprocities that establish fundamental norms for human virtue or vice, and they might, as some focus group members did, see manipulations undertaken strictly for personal amusement as deeply troubling.

But agrarians will baulk at the suggestion that the profit motive also signals moral trouble. In arguing for restraint, Fiester claims that alteration of animals is a comparatively new thing and that genetic transformations are especially troubling in virtue of being irreversible. It is, of course, possible that long-standing agricultural breeding practices controvert these points, but other claims Fiester makes would, in any case, appear to make ordinary farm production of animals unethical. She leaves open the possibility that compelling medical or veterinary applications of the technology may be capable of overcoming her presumption of restraint. But stipulating a moral spectrum for animal biotechnology with these simple 'yes' and 'no' endpoints fails to provide adequate guidance for animal biotechnology applications in the domain of agricultural ethics, where even a Hesiod or a Wendell Berry will have the sense to see that farmers tend their animals because they hope to profit from doing so.

Agricultural animal breeding has always been undertaken for purposes of economic advantage. First ordinary farmers and later breeding specialists selected for traits that would obtain higher prices in livestock sale barns, or that would increase the profitability of animal husbandry. Such traits have included heavy muscling, especially in carcass areas that produce valuable cuts of meat, or have been focused on efficient feed conversion and rapid growth. In recent decades, the toll of these breeding practices on the animals themselves has become apparent. High-producing dairy cows are subject to higher rates of mastitis infection. Heavily muscled animals of many breeds experience skeletal deformities, or may be incapable of unassisted reproduction. Although it would be easy to overstate the extent to which animal welfare has been recognized as a legitimate issue within farm animal research and production today, many believe that genetic adjustments will be needed to return animal husbandry to acceptable norms of humane treatment (Gamborg and Sandøe 2003). While it is certainly plausible to hold that some of these extreme problems associated with animal genetics are grounded in an industrial philosophy of agriculture, all farmers, agrarian and industrial alike, are raising animals with some notion of return on their labour residing somewhere in their minds.

The animal welfare problems just noted make it clear that science focused exclusively on the profitability of agriculture can become morally problematic. Yet agricultural researchers and their constituency have more typically viewed science that results in production-enhancing technology as a public mandate, rather than as something to be ashamed of. If we have now come to see not only certain genetic technologies, but also many pesticides and synthetic fertilizers as flawed, it nevertheless distorts both the motives of researchers and the culture within agricultural research contexts to lay much stress on the fact that such agricultural research is geared towards economic benefit. Although this research was indeed dedicated towards making farmers more profitable, it was untypical for researchers themselves to reap economic rewards beyond the salaries they drew from publicly funded universities and government labs. This is the culture in which a great deal of the capability for animal biotechnology has been developed. In this context, working to benefit historically (but no longer) poor and politically weak farm families has created a legacy that still marries biblical and democratic mandates in the minds of many practitioners. This mode of thinking manifests itself in the assumption that productivity-enhancing work will eventually benefit the poor, if not here, then perhaps in the developing world.

Although it is easy to look on contemporary animal science, including biotechnology, as thoroughly intertwined with the industrial philosophy of agriculture, it is nonetheless still the case that many of the individuals who undertake this research think of themselves in agrarian terms. Even those who emphasize the benefits of productivity enhancement, echoing the words of Jeffery Sachs, may be envisioning their science as benefiting people whom they think of more along the model of an agrarian yeoman than a desktop farmer, anxious to finish the day's hedge trades so he can get to the golf course. And benefits to the poor, so ably articulated in the industrial philosophy, are not to be sneezed at, in any case. It has not, therefore, surprised me when I have discovered agrarian hearts beating beneath the breast of more than one high-powered genetic engineering or cloning scientist that I have chanced to meet in my career as an agricultural ethicist.

Michiel Korthals (2008) writes that while agricultural science was once construed as a battle against hunger, it has now become embroiled in a battle against alternatives. In this second battle, scientists have lined themselves up against a variety of sociocultural identities that people construct and maintain through their food practices. He goes on to argue that principles of procedural justice weigh in favour of ensuring that voices articulating these constructions are heard in the debate over new technology. Korthals's argument would apply to many of the concerns with animal biotechnology that have been articulated in the focus groups and analysed in other chapters of this book. What the opposition of industrial and agrarian philosophies of agriculture adds to his argument is a tradition of historically grounded ways in which people have thought that the structure and organization of agriculture mattered morally and socially. Certainly many of Korthals's sociocultural identities are wrapped up in agrarian ideals. Our ethical struggle consists in finding the proper ears with which to hear the voices that call for an agrarian world, and in understanding how to balance these calls in light of hunger and of social, environmental, and public health externalities.

References

Bauer, G. 2008. *Farm sanctuary: Changing hearts and minds about animals and food*. New York: Touchstone Books.

Berry, W. 1977. *The unsettling of America: Culture and agriculture*. San Francisco: Sierra Club Books.

Campbell, J. 2000. Franklin agrarius. In *The agrarian roots of pragmatism*, ed. P.B. Thompson and T.C. Hilde, 101–17. Nashville, TN: Vanderbilt University Press.

Donahue, B. 2003. The resettling of America. In *The essential agrarian reader*, ed. Norman Wirzba, 34–51. Lexington: University Press of Kentucky.

Fiester, Autumn. 2008. Justifying a presumption of restraint in animal biotechnology research. *American Journal of Bioethics* 8(6): 36–44.

Gamborg, C., and Sandøe, P. 2003. Breeding and biotechnology in farm animals. In *Key issues in bioethics: A guide for teachers*, ed. R. Levinson and M. Reiss, 133–42. London, UK: Routledge.

Goldberg, R., Rissler, J., Shand, H., and Hassebrook, C. 1990. *Biotechnology's bitter harvest*. Washington, DC: The Biotechnology Working Group.

Hanson, V.D. 1996. *Fields without dreams: Defending the agrarian ideal*. New York: The Free Press.

Holland, A. 1995. Artificial lives: Philosophical dimensions of farm animal biotechnology. In *Issues in agricultural bioethics*, ed. T.B. Mepham, G.A. Tucker, and J. Wiseman, 293–305. Nottingham, UK: University of Nottingham Press.

Kalter, R. 1985. The new biotech agriculture: Unforeseen economic consequences. *Issues in Science and Technology* 2: 125–33.

Korthals, M. 2008. Two battles in the history of agriculture: Against hunger and against alternatives. Comment on John Perkins' and Rachael Jamison's 'History, ethics and intensification in agriculture.' In *The ethics of intensification: Agricultural development and cultural change*, ed. P.B. Thompson, 91–6. Dordrecht, NL: Springer.

Lacy, W., Lacy, L., and Busch, L. 1992. Emerging trends, consequences and policy issues in agricultural biotechnology. In *Bovine somatotropin and emerging issues: An assessment*, ed. M.C. Hallberg, 3–32. Boulder, CO: Westview Press.

McKibben, B. 2007. *Deep economy: The wealth of communities and the durable future*. New York: Times Books.

Munro, W.A., and Schurman, R.A. 2008. Sustaining outrage: Cultural capital, strategic location and motivating sensibilities in the U.S. anti-genetic engineering movement. In *The fight over food: Producers, consumers and activists challenge the global food system*, ed. W. Wright and G. Middendorf, 145–76. University Park: Penn State University Press.

Nelson, S. 1998. *God and the land: The metaphysics of farming in Hesiod and Vergil*. Oxford: Oxford University Press.

Norton, B. 2005. *Sustainability: A philosophy of adaptive ecosystem management*. Chicago: University of Chicago Press.

NRC (National Research Council). 1989. *Alternative agriculture*. Washington, DC: National Academy Press.

Pollan, M. 2008. *In defense of food: An eater's manifesto*. New York: Penguin Books.

Sachs, J.D. 2006. *The end of poverty: Economic possibilities for our time*. New York: Penguin Books.

Shiva, V. 1991. *The violence of the Green Revolution*. London: Zed Books.

Smith, K.K. 2003. *Wendell Berry and the agrarian tradition: A common grace*. Lawrence: University of Kansas Press.

Thompson, P.B. 1997. Science policy and moral purity: The case of animal biotechnology. *Agriculture and Human Values* 14: 11–27.

– 2001. The reshaping of conventional farming: A North American perspective. *Journal of Agricultural and Environmental Ethics* 14: 217–29.
– 2007. Theorizing technological and institutional change: Alienability, rivalry and exclusion cost. *Technè* 11(1): 19–31.
Verhey, A. 1995. Playing God and invoking a perspective. *Journal of Medicine and Philosophy* 20: 347–64.
Wargo, J.P. 1996. Our children's toxic legacy: How science and law fail to protect us from pesticides. New Haven, CT: Yale University Press.
Wunderlich, Gene. 2000. Two on Jefferson's agrarianism. In *The agrarian roots of pragmatism*, ed. P.B. Thompson and T.C. Hilde, 254–68. Nashville, TN: Vanderbilt University Press.

6 The Health Researchers: Transformational Research in a Transactional World

LORRAINE SHEREMETA

Human use and manipulation of plants and animals for food and medical treatment is not new. Insulin from pigs has been used to treat diabetes in humans since the 1920s. Over the past fifty years, however, the use of animals in biomedical research has become a topic of increasing public attention and concern. Over the last thirty years, biomedical research involving the direct genetic manipulation of animals (and of humans) has become a lightning rod for intense debate. Test-tube babies created by in vitro fertilization and the successful cloning of Dolly the sheep and the spectre of human cloning that it raised continue to resonate in discussions about genetic technologies. The present debate about the use of genetically modified animals in research is but one piece of the broader societal debate (or set of debates) about genomics and about animal research in the broadest sense. The issues concerning genetically modified animals in biomedical research are complex and bound, in various ways, to these broader related discussions.

In contrast to the majority of authors in this book, I have focused on only one aspect of animal biotechnology – the direct manipulation of genes in the animal genome to produce genetic variations that could not be achieved by conventional means. I have not considered cloning technologies. While not dependent on direct intervention using transgenic techniques, selective breeding has been of significant importance to biomedical research. For example, it has led to the development of hundreds of inbred strains of genetically identical mice with characteristics that make them highly valued as research tools (Beck et al. 2000). Research using inbred strains of mice has resulted in several Nobel Prizes, which speaks to the importance of inbred strains to various facets of biomedical research (Beck et al. 2000; Manis 2007).

Recently, the ability to genetically modify animals in specific ways has led to an increase in the use of animals, most notably of mice and fish, in research. In the United Kingdom, for example, the use of transgenic animals in research has increased tenfold over the last decade (Watts 2007). This goes some way to explaining why the twenty-year trend in the reduction of animal use has stopped and appears to run counter to attempts to reduce the numbers of animals used in research, to refine research procedures to minimize animal suffering, and to replace animal research with other methods (Russell and Burch 1959; Kmietowicz 2002). Attempts to distinguish those issues that are specific to transgenic animal care and use are ongoing. To date, most attention has been directed to animal welfare issues, risks to humans who work with transgenic animals, and unique environmental risks that accidental or intentional release of transgenic animals into the environment could pose.

Europe has been, and continues to be, the focal point for discussion about the use of animals in research, though concerns are increasingly voiced by groups around the world (Brumfiel 2008; Huggett 2008). Concern is evidenced by the global proliferation of animal rights and animal welfare groups and of legislation governing the care and use of animals in research (Rollin 2006). It is also evidenced by the increasing number of illegal acts perpetrated by animal rights extremists against big pharma, animal research facilities, and researchers themselves, as well as in the passage of specific laws against animal extremism (Huggett 2008). The emergence of extreme opposition to animal research, the transnational nature of medical product development, and the reality that a global marketplace exists for such products compels a global dialogue about the appropriate use of animals in biomedical research. The dialogue is further compelled by the common criticism, founded or not, that animal research and testing is of limited value and that its contribution to the treatment of human diseases has been overstated (Pound et al. 2004).

The purpose of this chapter is to provide some background and rationale for and against the use of transgenic animals in industrially driven biomedical research and to consider how scientists and health researchers working in the field and animal welfare proponents view such research. Although the focus group discussions were broad in scope, much of the discussion proffered in this chapter naturally focuses on the use of genetically modified mouse models for industrially driven biomedical research. Discussion about the challenges, opportunities, and relevance of improving oversight mechanisms for animal research will be provided. Consultation and communication with interested groups, including scientists, consumers, medical professionals, farmers, and other interested groups as part of this process is of critical importance.

Why Is Animal Research Necessary?

Ethical Norms, National Laws, and International Standards Demand It

The ethical and legal requirement for biomedical research to be performed in animals prior to human investigation (preclinical testing) emerged in the aftermath of the Second World War. Following the war, the Nuremberg Trials considered the atrocities of human experimentation by the German Nazi regime. The Nuremberg Code (1947) formalized research to encompass systematic study, unprocurable by other methods, to benefit society and required that, where appropriate, animal experimentation should precede human studies (Freyhofer 2004). The requirement for animal studies has been further entrenched in articles 11 and 12 of the World Medical Association's Declaration of Helsinki (World Medical Association 1964), as follows:

> Article 11 – Medical research involving human subjects must conform to generally accepted scientific principles, be based on a thorough knowledge of the scientific literature, other relevant sources of information, and on adequate laboratory and, where appropriate, animal experimentation.
>
> Article 12 – Appropriate caution must be exercised in the conduct of research which may affect the environment, and the welfare of animals used for research must be respected.

The Declaration of Helsinki has been incorporated by reference in the ethics codes of professional medical organizations, pharmaceutical companies, and industry organizations around the world. It is specifically referenced in virtually all human clinical-trial protocols. At the national level, the requirement for preclinical animal studies in drug development is firmly entrenched in the law and is a standard requirement pursuant to the laws and regulations governing drugs, medical devices, natural health products, and veterinary products.

This declaration is acknowledged in the International Conference on Harmonization (ICH) *Good Clinical Practice* guidelines (ICH 1997). Good clinical practice (GCP) is an internationally recognized ethical and scientific standard for the conduct of clinical trials. Adherence with GCP ensures adherence to the principles originating in the Declaration of Helsinki. It also facilitates experimental consistency and high-quality data as agreed by the regulatory agencies of the European Union, Japan, and the United States. Health Canada has formally adopted the GCP guidelines.

Scientific Progress and Innovative Medical Product Development Require It

Animal research has been, and continues to be, viewed by many, especially those in the biological sciences and the health and medical professions, as necessary for the accrual of basic knowledge that will inform progress in these areas. Basic research involving transgenic animals has moved to the fore as significant efforts have been directed towards the understanding of gene function in the aftermath of the publication of the human genome sequence (Lander et al. 2001). At the time of publication of the draft sequence, the researchers were quick to point out that the sequence data should be viewed as a 'launch pad for new discoveries, not an end in itself' (Sanger Centre website).

Since the completion of the Human Genome Project sequencing effort, functional genomics and comparative genomics have emerged as major research endeavours. The Human Genome Project was the first major step in understanding humans at the molecular level. Though the sequencing is complete, many questions remain unanswered. Fundamental knowledge about the function of most of the estimated 30,000 human genes has yet to be assembled and exploited. To that end, functional genomics focuses on understanding the function of genes and other parts of the genome. Comparative genomics is the analysis and comparison of genomes from different species with the purpose of gaining an understanding of how species have evolved and to determine the function of genes and non-coding regions of the genome.

It has been known for some time that many biological processes are shared by many species of animals, including humans. Researchers have learned a great deal about the function of human genes by examining their counterparts in simpler model organisms, such as the mouse.

In 2002, the genetic sequence of the laboratory mouse was published (Waterston et al. 2002). The remarkable similarities identified between the human and mouse genomes have inspired much speculation about the relevance of gene-function studies performed in mice to the understanding of human disease. Comparative analysis of the human and mouse genomes, for example, reveals that greater than 80 per cent of the mouse genes have the same functions as in humans. The human and mouse genomes are of comparable size, they encode for a similar number of genes, and for the most part, the same genes are clustered on equivalent chromosomes. These findings have resulted in much higher priority being placed on the development and characterization of new strains of genetically modified mice to facilitate a better understanding of human physiology and disease. At present, over 95 per cent of transgenic animals used in biomedical research are mice (United Kingdom Home Office 2006; Canadian Council on Animal Care 2006). A cursory search of the International Mouse Strain Resource, a publicly accessible online database of mice used for research purposes, retrieved over 2000 strains of transgenic mice (International Mouse Strain Resource website).

Animal Research and the Innovative Development of New Biomedical Products

The standard drug development pathway spans basic research, including the identification of drug targets, preclinical research (in vitro and animal testing), clinical research (phases I–IV) to define safety, effectiveness, and long-term-use data (figure 1). The cost of getting one new medicine to market is in the USD $1 billion range and the process typically takes ten to fifteen years. While the number of drug approvals has remained relatively constant over time, the annual investment into research and development has risen dramatically – from approximately USD $1 billion in 1975 to $40 billion in 2003 (Erickson 2006). It has been estimated that of five to ten thousand potential drug candidates, only one will ultimately be licensed as a drug product (Pharmaceutical Research and Manufacturers of America [PhRMA] 2007). The number of animals, both transgenic and other, used over the entire drug development process is difficult to quantify. To further complicate matters, the numbers of animals required as part of the regulatory

Figure 6.1: The life of a medicine (modified from Nuffield Council on Bioethics 2005: 146, with permission)

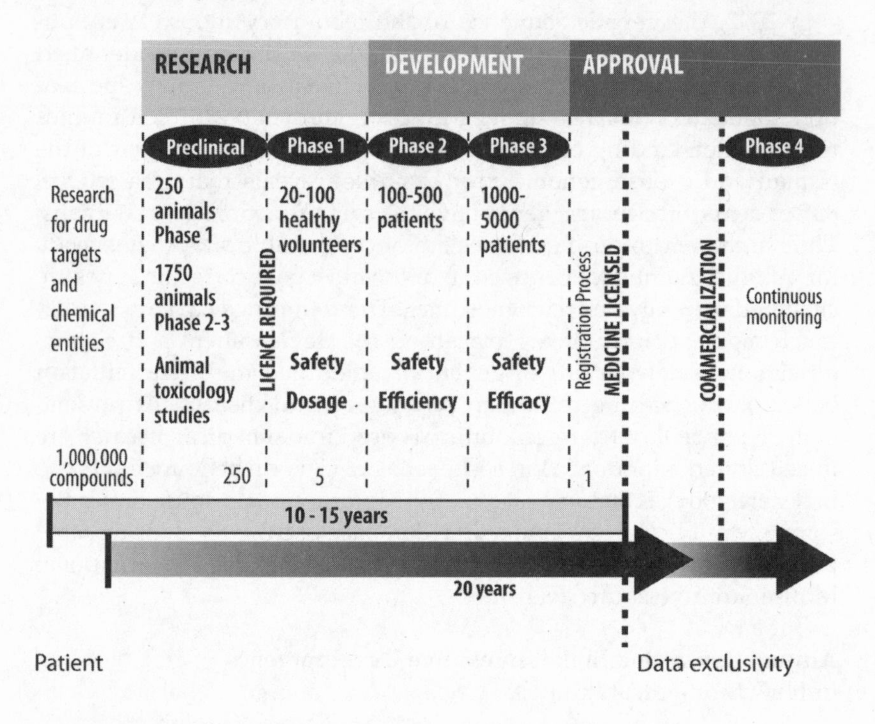

approval process is not clearly defined. By one estimation, for each novel drug target, approximately 2000 animals will be used in early preclinical biodistribution and toxicological testing (Nuffield Council on Bioethics 2005: 142).

The selection of drug targets to treat human disease has been described as evolving 'from managed serendipity to engineered selection' (Pritchard et al. 2003: 542). New approaches based on genomics, proteomics, metabolomics, and combinatorial chemistry are fuelling an expansive pipeline of potential new targets (Zambrowicz and Sands 2003). The critical determination of when a new chemical entity can be administered to humans for the first time depends almost entirely on the results of preclinical studies performed in animals in order to predict the behaviour of the compound in humans and to evaluate the potential toxic effects. Later in the drug development process, long-term safety studies in animals are needed before a licence to market the drug product can be granted (Pritchard et al. 2003).

Of late, the US Food and Drug Administration and regulatory authorities around the world have been criticized for their inability to effectively oversee the safety of drug products reaching the marketplace (US FDA, Subcommittee on Science and Technology 2007). The recent controversy over the approvals and subsequent voluntary withdrawals of the two Cox-2 inhibitors Vioxx and Bextra has significantly undermined trust in the regulatory process (US FDA 2005; Psaty and Charo 2007) and in drug companies seeking to get their products approved. It is argued that regulators need more and better knowledge about the characteristics of potential new drugs so that they can make better decisions as to whether or not to approve them. The key is to be able to predict accurately, or to know what data will give the best prediction of, safety and efficacy in humans. Questions remain: How can animal biotechnology be used to generate more robust data for regulators? Are there other, non-animal, approaches that may be employed to generate better data for regulators?

How Are Transgenic Animals Currently Used in Biomedical Research?

Given the history of medical research, the emergence of bioethics as a formal discipline, and the likely relevance of animal models to human disease, the use of animals has become fundamental to all aspects of biomedical research.

Transgenic animals are used in a variety of ways in biomedical research. The following section provides a summary of the various uses, and potential uses, of genetically modified animals in relation to the development of new medicines.

- *As models of human disease processes:* Mice and other animals can be specifically modified through the introduction or inactivation of specific genes that are chosen to help researchers to understand the cause and manifestation of human diseases. Human diseases that arise from a mutation in a single gene can be readily modelled in the mouse, and specific genetic modifications can be introduced at will.
- *To understand gene function:* The Human Genome Project is a vast international effort that has facilitated the sequencing and

identification of some 30,000–40,000 genes that make us human. The critical next step is to understand how these genes function. Knowledge in this regard is still very limited. Given the striking similarities between the mouse and the human genome, it is expected that by systematically altering ('knocking out' or 'knocking in') a gene, causing them to be inactivated or expressed and studying the effects of the change, researchers can learn about the function of genes in humans. These mice are then bred, creating populations of offspring with the trait.

- *To understand developmental biology:* Genetic modification of several species including Drosophila (fruit-fly), Brachydaniorerio (zebrafish), and the Xenopus (frog) are proving useful to understand basic developmental biology. Zebrafish have become a major model system because of the practical ease with which they can be used to identify the function of novel genes.
- *Drug target identification:* A person's genetic make-up is important for understanding how diseases behave. It also plays an important role in the response to drugs. Identifying and understanding how certain genes modify disease genes through the use of animal models helps researchers understand disease processes and is useful in identifying useful targets for new treatments.
- *Safety and efficacy testing of drug candidates:* It is expected that the efficacy of a drug candidate can be best tested in a genetically modified animal model that most closely mimics the human disease that the drug is designed to treat. It is hoped that transgenic models will be valuable to demonstrate the potential side effects of drug products that are tailored to influence gene expression. Currently, both traditional and genetically modified animal models are used in these processes to ensure that only those drugs likely to be both safe and effective will proceed to investigation in human subjects.
- *Toxicity testing:* Currently, genetically modified animals can be used to test chemicals and drugs to ensure they are not carcinogenic. They are also used to some extent to test for other mechanisms of toxicity and for embryonic damage. It is hoped that animals with specific mutations can dramatically speed up the time required for genotoxicity testing. Importantly, mice have been modified to act as test systems for specific effects that could previously only be tested on higher primates.
- *For production of therapeutic proteins:* Transgenic animals can be engineered to act as 'bioreactors' for biological products that can be

used for medical treatment of humans. Animals can be engineered to produce proteins in their milk or other tissues. In 2006, the European Commission approved GTC Biotherapeutics' drug ATryn, an anti-clotting protein produced in the milk of transgenic goats. The drug is approved for the treatment of patients with hereditary antithrombin deficiency that are undergoing high-risk surgical procedures or childbirth (GTC Biotherapeutics website). In September 2008, the United States Food and Drug Administration assigned a priority review to ATryn. GTC maintains a herd of 1400 goats in Charlton, Massachusetts, as a living drug factory (Lei 2006).
- *For production of humanized organs for transplantation (Xenotransplantation):* Transgenic animals with human genes can be bred so that their organs are sufficiently 'human' to prevent their rejection by the immune systems of organ recipients.

It is important to note that while the uses of transgenic animals described above focus on the development of new treatments for human disease, it is hoped and expected that better understanding of gene function and disease processes will have spinoff benefits for veterinary medicine and animal health (Kling 2007). It is also hoped that by opening the markets for companion animals and veterinary products the animal biotechnology sector will attract significant investment to the field.

The Case for Animal Research – the Industrial Perspective

The industrial perspective on biotechnology is presented in this section as a composite of the views and policy positions taken by the Biotechnology Industry Organization (BIO), GlaxoSmithKline (GSK), and Novartis. Each is included on the basis of their profile and role in the commercial development of biotechnology. According to their communications materials, the motivations of BIO and of its constituent members, which include GlaxoSmithKline and Novartis, are aligned yet distinct. BIO is an industry organization representing more than 1150 biotechnology firms, academic institutions, centres, and related institutions within the United States and in thirty-three countries around the world. Members of BIO are diverse and participate in the research and development of bioproducts that span economic sectors including agriculture, human and veterinary health, and the environment.

Given that the development of innovative biomedical products depends on biotechnology and other enabling technologies, BIO argues that its members' voices are important when considering animal biotechnology. It is also important that companies are represented given that, to date, they have been front and centre in the broader societal debates: Monsanto in the case of genetically modified crops and Myriad Genetics in the case of hereditary breast-cancer genetic testing. There exists a deep scepticism as to the profit motivation of drug companies and the degree to which that motivation can and does align with the public good (Caulfield et al. 2003).

Battles over the ownership and control of genetic technologies have polarized the discussion and have entrenched 'good' and 'bad' positions that are presumed inherent in the not-for-profit/public and for-profit/private spheres. As a case in point, *Harvard College v. Canada (Commissioner of Patents)*, a patent dispute that landed in the Supreme Court of Canada, is telling. The case arose over the patentability of transgenic mice specifically tailored to be susceptible to cancer. The case raised issues about the patentability of higher life forms and the commercialization and commodification of life. The court's decision was a technical, legal decision: the mouse, as a modified higher life form, is not patentable. The patentability of the process used to create the transgenic mouse was not at issue. Most non-lawyers viewed the decision as a victory for the 'public' and a loss for the 'private.' In actual fact, the decision was a victory for both sides. On the face of it, Canada's patenting position in the international community appears nuanced and ethical. However, whether you can patent the mouse or the process to create the mouse is, for the most part, a legal distinction without a practical difference. In effect, transgenic mice and other non-human animals can be monopolized in Canada (as in other jurisdictions) if the method to create them satisfies the legal criteria for patentability. Whether justified or not, general concerns about the patenting and commercialization of new drugs and medical technologies persist. In fact, related concerns arose in the discussions of each of the focus groups considered in this chapter.

The shift towards managed corporate social responsibility and ethical business practices over the last twenty-five years or so has been and continues to be an important consideration for biotechnology firms. The green movement has shown that success in the marketplace is increasingly dependent on recognition of and alignment with societal goals. The creation and commitment of investors to green investment funds has led to improved practices and accountability. For drug companies, animal use is an area of particular concern. Overall, it is apparent from their corporate messaging that BIO, GSK, and Novartis support biotechnology and transgenic technologies, but not at any cost. They recognize the value of science and the need to treat animals humanely if they are to get data that will support product registration.

The degree to which corporate messaging can be trusted is yet another matter. It has recently been reported that public trust in bioscience firms is at an all-time low, with only 13 per cent of Americans believing that the industry is honest and trustworthy (down from 80 per cent in 1997) (Finegold and Moser 2006). Purportedly, the distrust has arisen from a variety of causes, including several high-profile cases involving the withholding of evidence concerning adverse effects of drugs, falsification of research data, and questionable marketing practices by drug companies. The tendency to treat all drug companies as bad corporate citizens is likely an unjust oversimplification of reality. The perception that industry is bad while academia is good fails to recognize the frequency with which individuals have activities and interests that traverse both spheres.

The following section summarizes the relative positions of the Biotechnology Industry Organization, GlaxoSmithKline, and Novartis with respect to transgenic animal research as reported through their respective corporate messages on their websites.

The Biotechnology Industry Organization (BIO)

As noted, transgenic animals are increasingly used in the drug discovery and development process. The ultimate goal of biomedical research is to find ways to prevent disease or to develop better therapeutic options to treat disease processes. Transgenic animals are used in various ways to help researchers understand the role of genes in disease processes. This is important as the pharmaceutical industry is under extreme pressure to develop more and better drug products in the wake of knowledge gleaned from the full sequencing of the human genome. To date, the expectations raised in the wake of the genomic era have not been met. In fact, the number of drug products approved by the United States FDA and regulatory agencies around the world has levelled off over the past few years (PhRMA 2008).

Generally, animal research is perceived to be of fundamental importance for future developments in biomedicine. BIO, the world's largest biotechnology industry organization, claims that biotech companies have relied successfully on animal research 'to develop more than 160 drugs and vaccines approved by the US Food and Drug Administration, helping 325 million people worldwide and preventing incalculable human suffering' (BIO website). Despite this optimistic outlook, significant challenges remain in understanding many serious human diseases and conditions that afflict humans (and animals).

In a recent report, BIO argues that the development of agricultural and medical applications of biotechnology will have profound benefits for society and that 'the needs for public health and security are urgent' (Gottlieb and Wheeler 2008: 35). GM animals enable the production of safer and lower-cost proteins and drugs that have the potential to transform medical practice and to

> deliver substantial improvements in terms of cost, safety and availability of urgently needed drugs and treatments, bringing substantial health benefits. Likewise, genetically engineered animals can also sustainably, and in an environmentally friendly and pro-welfare friendly manner, meet the growing global demand for high quality and safe animal food products. (ibid.)

BIO recognizes that for anyone seeking to develop new biotechnology-enabled medical products, animal research is required by regulatory agencies as part of the development process if that product is to be marked for human or veterinary use. The organization anticipates that transgenic technologies will lead to the discovery of novel drugs and therapies, including gene therapy and cellular therapies. In the end, it suggests that uncertainty in law, regulation, and policy is hindering the timely development of needed biotechnology products and that clarification in these areas is urgently required.

That BIO is actively involved and interested in various ethical, legal, and social issues is clear from its website. Its Statement of Ethical Principles for the Care and Use of Animals in Biotechnology Research is a testament to this fact (BIO website). This statement is important in that it conveys industry's recognition that animal biotechnology is important to the success of its members and to the improvement of human and animal life; but so is the humane treatment of animals. BIO expressly advocates that, whenever possible, research methodologies that reduce, replace, or refine the use of animals in research should be used. It also is of the opinion that biotechnology offers solutions to further reduce the number of animals in research, although it's not clear what shape these solutions might take.

BIO recognizes the desirability of setting and maintaining high standards of care for animals used in biotechnology research. It refers specifically to guidance documents published by the Institute for Laboratory Animal Research, Commission on Life Sciences, National Research Council (*Guide for the Care and Use of Laboratory Animals*, 7th ed., 1996), and the Federation of Animal Science Societies (*Guide for the Care and Use of Agricultural Animals in Agricultural Research and Teaching*, 1999). BIO undertakes to work with the relevant regulatory agencies to ensure high standards of care and use for all animals used in biotechnology research.

With regard to public awareness about biotechnology research, BIO recognizes that there is a need to inform the public about how biotechnology research involving animals is being applied in human health, animal health, agricultural, industrial, and environmental areas. It claims to recognize the importance of open communication and engagement with academics, consumers, medical professionals, farmers, lawmakers, scientists, and other interested groups.

GlaxoSmithKline

GlaxoSmithKline (GSK), one of the world's five largest pharmaceutical companies, reports on its website that through commitment to the 3Rs (reduce, replace, or refine), despite a significant increase in research and development activity since 1994, the use of animals in GSK research has not increased proportionately. GSK claims that, as part of its business model, the use of transgenic animals is helping in the discovery and development of new treatments and cures for disease. It argues that the use of transgenic mice is important in that it will contribute to the further reduction in the use of higher animals, including dogs and primates, in biomedical research (GSK website).

Like BIO, GSK appears to hold the general position that animal welfare issues associated with the use of transgenic animals are fundamentally no different from those arising with other animals in the course of biomedical research. It claims that of primary importance is the need to minimize pain or distress to individual animals that are used in research and not the manner in which the animals are bred or their genetic make-up.

GSK claims that the development and subsequent use of transgenic animals is subject to stringent internal review and government regulation and oversight. As noted previously, regulations that control animal research permit the use of animals only when no appropriate and validated alternative exists. GSK recognizes that there is public concern over the use of transgenic animals in research and argues that it is committed to addressing these concerns.

In conclusion, it is the position of GSK that

> gene-based biomedical research offers one of the best hopes yet for curing the major diseases which still afflict mankind. The use of transgenic animals is central to realizing that hope and offers the potential for the use of fewer animals in more targeted experiments. We must be clear. There are only two alternatives to using animals. One is to use humans in basic research; the other is to delay or even give up the search for desperately needed new treatments and cures. (GSK website)

GSK views the use of transgenic animals as necessary, with the potential for significant human health benefits. It lays down the challenge for governments, industry, and society to collectively ensure that 'transgenic research continues to be sensitively carried out for proper medical ends in a suitably balanced regulatory environment' (GSK website).

Novartis

Novartis has an Animal Welfare Policy that sets out key principles, requirements, and responsibilities relating to animal welfare in animal research that is performed in-house or that is contracted out by Novartis (Novartis website). Specifically, the policy states that all animals used in research must be treated humanely and cared for in accordance with the particular needs of the given species and affirms Novartis's commitment to the 3Rs.

Novartis claims that animal testing is a core element in pharmaceutical research and development and that it enforces the highest standards of animal welfare and good scientific and ethical practice. It argues that where animal testing is the only method available, a careful review of the experimental procedure and rationale for the selection of a specific animal species and definition of the test population is performed. According to the website, where external contractors are needed to carry out specific research, they are carefully chosen and audited to ensure that they meet or exceed the animal welfare standards set by Novartis.

Novartis argues that all animal research done by or for Novartis must be compliant with relevant legislation dealing with animal welfare and experimentation in the jurisdictions where it is performed (Novartis website). Where research is performed in less regulated countries, Novartis requires that its own standards are met and that all animal studies are performed to meet the requirements of United States law and regulation.

The Novartis website states that it is currently using transgenic animals to obtain new information on diabetes and Alzheimer's disease. It also states that the ethical and welfare standards for the use of transgenic animals are the same as those used for animals that are not transgenic. Novartis claims to be aware that the use of transgenic animals in pharmaceutical research raises ethical and moral concerns and that these concerns must be openly discussed. It specifically notes environmental and biosafety concerns and impacts on the environment, as well as the morality of crossing species barriers and patenting life. Like BIO and GSK, Novartis welcomes responsible public debate on biotechnology, its applications, and the benefits it provides (Novartis website).

The Case against Animal Research – Science-based Concerns and Animal Welfare

For some, the role of animal research in the development of new medical treatments for complex human disorders raises ethical concerns. On the one hand, organizations like BIO and its drug company members argue that research involving animals holds promise for aiding humanity. Opponents argue that animal research is largely unhelpful in the development of treatments for humans and that non-animal alternatives could eliminate the need for animal research altogether. The late David Horrobin, renowned researcher and entrepreneur, noted:

> Congruence between in vitro and animal models of disease and the corresponding human condition is a fundamental assumption of much

> biomedical research, but it is one that is rarely critically assessed. In the absence of such critical assessment, the assumption of congruence may be invalid for most models. Much more open discussion of this issue is required if biomedical research is to be clinically productive. (Horrobin 2003: 151)

Horrobin questions whether researchers wouldn't do better to focus on humans to understand human disease. In his opinion, such efforts might be 'at least as productive as the massive investment in the investigation of unvalidated animal and in vitro models' (ibid.: 153). Contrary to the optimism espoused by BIO, Horrobin suggests that pharmaceutical research is failing in its ability to deliver new drugs and that research has taken a 'wrong turn in its relationship to human disease' (151).

Other commentators have noted that inadequate animal models are a major hurdle in the overall process of drug development (Wall and Shani 2008; Hackam and Redelmeier 2006). Data suggest that the use of animal models is an inexact science, but for now, computer models, in vitro systems, and lower-animal models are of limited use. Animal models remain the best alterative in many situations. It is widely recognized that the development and validation of better alternatives to animal models is needed (Balls and Fentem 1999; European Partnership for Alternative Approaches to Animal Testing 2007). Efforts to promote the adoption of new, non-animal, test methods are under way in Europe (European Centre for the Validation of Alternative Test Methods), Germany (Zebet – Centre for Documentation and Evaluation of Alternatives to Animal Experiments), and the United States (Interagency Coordinating Committee on the Validation of Alternative Methods). These entities reflect the global and governmental recognition of the need for alternatives and the development of better regulatory testing schemes for new materials and consumer products (Kroeger 2006).

Fund for the Replacement of Animals in Medical Research (FRAME)

FRAME is a registered charity in the United Kingdom, founded in 1969 to promote the development and adoption of methods of experimentation that can be used instead of animal experimentation. Its ultimate objective is the elimination of animal research; FRAME acknowledges, however, that 'the immediate elimination of all animal research is not possible' (FRAME website). FRAME acknowledges the importance of medical research for the development of new treatments for human and animal diseases. It also recognizes that new consumer products, medicines, and chemicals must be tested 'to identify potential hazards to human and animal health, and to the environment' (ibid.).

FRAME is concerned about the marked increase in the number of genetically modified (GM) animals used in scientific procedures in the United Kingdom. Statistics from the Home Office in that country show that the number of GM animals used in research is increasing, and that by 2005, procedures using GM animals represented 33 per cent of all procedures (UK Home Office 2006). The largest proportion of those procedures involves GM mice (48 per cent), with the most common purpose for such procedures being breeding (ibid.).

Given the trend, FRAME expects to see significant increases in the numbers of GM animals used for the foreseeable future. In its view, this trend threatens to reverse the steady decline in the total number of procedures seen since the Animal (Scientific Procedures) Act 1986 came into force in Britain in 1987. The majority of GM animals used are mice, but as this technology is now increasingly being applied to other species, such as rats and fish, dramatic increases in uses of these species are expected (FRAME, Office-based research, http://www.frame.org.uk/page.php?pg_id=232).

FRAME is concerned about the impact of research with transgenic animals. It notes that the main purpose of procedures on GM animals has been for fundamental biological research, but the tangible benefits of such research are not yet readily apparent. It wonders at what point the benefits will become apparent.

FRAME is also concerned about the welfare of GM animals used for research purposes. FRAME argues that GM animals are often immunocompromised and susceptible to infection, and that animals bred as GM models of human disease may suffer as they develop disease. In addition, genetic modification of animals can result in unexpected and uncontrollable phenotypes. It may lead to welfare problems and suffering associated with genetic modifications that are not expected (FRAME, office-based research). As more and more animals are manipulated genetically, it seems reasonable to expect that the incidence of welfare problems will increase.

Unlike some animal rights groups, FRAME recognizes that there may be substantial benefits from using GM animals in research. Disease mechanisms may be better understood, new therapies may result, and eventually the number of animals needed for some research procedures may be reduced. However, FRAME considers that the existing regulatory controls for animal research are unsatisfactory in the context of GM animals and that further refinement of oversight mechanisms for this specific area is needed.

The Impact of Transgenic Techniques on Animal Use Data – Canada and the United Kingdom

Reports of animal use statistics from the United Kingdom prepared and presented by the Home Office on its Science, Research and Statistics website show that there has been a rise in the number of animals used in research, the underlying reason being the increasing use and reliance on transgenic animals for use as model systems. The same is true in Canada (Griffin and Gauthier 2007). Having said this, the current reporting system on animal use in Canada does not permit a clear accounting for the use of transgenic animals. As of 2006, the UK data show that transgenic animals are used in more than one third of all procedures in that country (UK Home Office 2006). Predictions suggest that large-scale functional genomics initiatives and the increasing demand for humanized transgenic animals will result in the continued increase in the numbers of transgenic animals used in research (Hudson 2007).

From analysis of the UK data, it is clear that fundamental research is causing a dramatic rise in the number of procedures performed and of animals used. By its nature, fundamental research is exploratory and highly specific to a given research question. It is, therefore, difficult and costly to develop non-animal alternatives. In this context, it is important to ensure that the fewest possible number of animals are used. One commentator notes: 'Despite the lack of systemic evidence for its effectiveness, basic animal research in the United Kingdom receives much more funding than clinical research. Given this and because the public accepts animal research only on the assumption that it benefits humans, the clinical relevance of animal experiments needs urgent clarification' (Pound et al. 2004: 514).

Several ways to reduce animal use have been suggested by the UK Home Office; they include:

- *Data sharing:* where possible, data should be shared to minimize repetition of animal experiments and tests.

- *Early adoption of new non-animal methods:* new test methods that will reduce the number of animals required in research and testing should be adopted as quickly as possible.
- *Use of experimental design and statistical clarity to minimize the number of animals used in research:* researchers and animal use committees should pay close attention to experimental design and statistical justification for the numbers of animals used in research and testing to ensure that the fewest possible animals are used (Hudson 2007).

From a practical perspective, the collection and presentation of animal use statistics should be harmonized across countries so that meaningful comparisons can be made and important trends in the data can be spotted early (UK Home Office 2006; Griffin and Gauthier 2007; Hudson 2007).

What Do Stakeholders Think? Issues Emerging from the Focus Group

In support of this volume, several focus group sessions were held, with the purpose of better characterizing the debate about animal biotechnology. Upon review of all transcripts, three were of particular relevance to this chapter and have been reviewed in detail. These include the Scientists, Health Researchers, and Animal Justice focus groups. Within the transcripts, there are differences but also a surprising degree of concordance.

The main substantive difference between the transcripts analysed is that participants in the Animal Justice group generally do not believe that animal research is useful – or at least they view it as not nearly as useful as researchers and drug companies would have us believe. For this group, there is little or no expectation that the current methods using genetically modified animals in research have the potential to transform modern medicine as proponents suggest. In addition, the costs are simply not worth the benefits, and there is a prima facie assumption that animal research is unethical. One participant in the Animal Justice focus group forcefully stated that 'animals are involved with this to their detriment . . . [and] I'm not in favour of any animal being used if it's to its physical or mental detriment' (Animal Justice stakeholder V1).

In addition, individuals in the Animal Justice focus group did not focus on the relative position of an animal in the phylogenetic scale to rationalize and accept research on some animals and not others. Though the improvement of human health is viewed as a laudable objective and one that should be pursued, it should be done without the use of animals. To rectify the current situation, an appropriate focus on alternatives is needed. This includes the development of non-animal research and test methods to ensure drug (and consumer product) safety. There was also a strong suggestion that if we need to know how certain things affect humans, humans rather than irrelevant animal models should be used as research subjects.

Generally, there is a strong sense that if people knew what was really going on in the world of animal research, whether transgenic or otherwise, they would not support it. In fact, to this group, the distinction between transgenic and non-transgenic animal use was largely irrelevant. One participant argued that, over time, our society has developed an unfounded 'predisposition' or 'tolerance' for animal research. It was aptly pointed out during the discussion that the significant headway made in reducing the number of animals used in research has been rapidly reversed by the increase in the number of animals used in transgenic research in recent years. For this group, democratic debate is welcomed; improved governance is clearly needed.

By way of comparison, many participants from the Scientists and Health Researchers focus groups do not 'like' animal research per se, and they clearly prefer not to have to perform animal research if they don't have to. They believe, however, that of the various tools available at the present time, animal research, and particularly research involving transgenic animals, is the best and perhaps only way to gain the necessary understanding of basic biological processes and complex human and animal diseases to enable the development of new approaches to treating human disease. They also believe that the current regulatory requirements for licensing drugs and medical devices demand research and testing using animals.

These participants are more inclined, though not universally so, to believe that research involving transgenic animals has the potential to transform medicine. While there was general acceptance of the current need for animal research, there was a simultaneous and contrary recognition that the system stands to be improved. They saw a need for the development of alternative approaches to research and testing, and there is hope that in future, as a result of research using transgenic animals, that fewer animals will be used in research and testing.

All three groups express clear concern that, at present, too many animals are being used for research and that unnecessary pain and suffering of animals, regardless of species, is not acceptable. The critical issue is that participants from the Animal Justice focus group were generally of the view that any research that precludes the animal living as nature intended is tantamount to unnecessary suffering. Given that views on pain and suffering are subjective, general agreement on this point is unlikely.

In both the Scientists and Health Researchers focus group transcripts, tones of both duty and utility emanate from the pages. One participant from the Scientists focus group agreed that humans have a responsibility to other animals on the planet. Animals are not just instruments to be used to benefit humans. There was agreement across the board that if animals are used in research, humans should ensure that animals have a potential to benefit; it should not be yet another case of animals used 'in the service of humans' (Health Researcher V1). When weighing the merits of potential applications that may be derived from biotechnology, a decisively utilitarian approach was articulated by one participant of the Health Researchers focus group: the greatest good for the greatest number is what we should strive for.

The issue of equitable distribution of the benefits of research was raised in various ways by the Animal Justice, Scientists, and Health Researchers focus groups. Concerns were raised in the Animal Justice group over the relative benefits ensuing that would accrue to the developed world and the developing world respectively. Though all groups recognized that monetary benefits would (or could) flow to companies selling products arising from biotechnology research, one participant from the Health Researchers focus group suggested that there was a potential for better access to drugs and an improved health care system. One participant from the Scientists focus group suggested that benefits to society, to patients, and to the developing world could result.

With respect to the commercialization of research, the Animal Justice focus group was the most sceptical of the motivations of drug companies, especially 'multinationals' and the capitalist ethos that they espouse. Concerns were raised about patenting and control in the biotechnology field and the questionable impact of such practices. One participant from the Scientists focus group noted that the agenda of companies may not necessarily accord with the broader societal agenda. Across all groups, it is evident that applications perceived to have been created out of industrial interest for monetary gain or that were perceived as frivolous were ranked relatively low. Conversely, applications with a potential health benefit for a great number of people (and animals) were ranked highly by most groups. Ironically, these are the very types of applications that are of highest priority to companies undertaking research in the area.

Review of the same three focus group transcripts from the perspective of recommendations for reform reveals an overall alignment of sentiment. Table 6.1 provides a synthesis of key suggestions put forth by participants in the Animal Justice, Scientists, and Health Researchers focus groups.

Arguably, the key suggestions described above reveal a remarkable alignment of priority and opinion between participants in the three groups. A fundamental question remains as to how animal research is governed and what are the principles and practices that are in place to ensure that genetically modified animals are used and cared for appropriately. The following section attempts to answer these questions from the Canadian perspective. Though not identical, the Canadian system is substantively reflective of systems in the United Kingdom and the United States, particularly with regard to the animal care review process. What will become clear upon review of the following section is that most of the suggestions raised by all three of the focus groups are expressly considered in existing oversight mechanisms in Canada. It has been reported elsewhere that the UK public would approve of current animal care and oversight mechanisms if they knew more about them (Festing and Wilkinson 2007a). After consideration of focus group transcripts and existing oversight mechanisms of animal research, it is the position of this author that, in all likelihood, Canadians would, similarly, approve of existing oversight mechanisms if they knew more about them.

Oversight of Animal Research in Canada

While different national systems of oversight of animal research may appear, at first blush, to be diverse, there are many substantial commonalities. Responsibility for the authorization of researchers to perform animal research can be vested nationally, regionally, or locally, as in Canada. Canada is credited for having created the first oversight system based on local Animal Care Committee (ACC) review of research. Now, ACCs, in one form or another, form an essential

Table 6.1
Key suggestions for systematic improvement of transgenic animal research: Animal Justice, Scientists, and Health Researchers focus groups

Animal Justice	Scientists	Health Researchers
– Do not permit unnecessary animal suffering.	– Do not permit unnecessary animal suffering.	– Do not permit unnecessary animal suffering.
– Focus on the development of alternative (non-animal) research and testing methods.	– Focus on the development of alternative (non-animal) research and testing methods.	– Focus research on alternative (non-animal) research and testing methods (use plants or unicellular organisms if possible).
– Critically assess existing evidence on the impact of research involving animals on the improvement of human health.		– Use animal model only if there is a clear goal and no alternative method available; economic justification alone is insufficient to justify animal research.
– Where possible, use humans as model system for research and testing.		
– Improve governance of animal research.	– Improve governance of animal research (science-based regulation). – Use expert decision-makers and consultative process to improve the existing system.	– Improve governance of animal research (labelling is an area that demands more attention).
– Consult the public; provide them with information needed to make informed decisions about animal research.	– Provide accurate information to the public; focus on the science and not on the potential applications.	– Focus research in areas where the need is most compelling; seek greatest potential benefit for greatest number of humans and animals.
– Engage in democratic and open debate about animal research.		

part of animal research oversight systems around the world (Gauthier 2007).

In Canada, the legislative jurisdiction over animal welfare falls substantially to the provinces. The federal government has a limited ability to impact the field through its power relating to criminal health law and federal spending. The Criminal Code of Canada protects animals from cruelty, abuse, and neglect. The Health of Animals Act and its regulations aim to protect Canadian agricultural animals from infectious diseases that threaten the health of animals and people. The federal spending power enables the federal government to set standards to be met by recipients of federal money. Such standards may be imposed on provinces, corporations, or individuals pursuant to granting rules or contract. This is the precise mechanism that the federal government uses to compel recipients of federal grant funding or contractors to comply with the standards set by the Canadian Council of Animal Care (CCAC). In addition, several Canadian provinces, including Alberta, Manitoba, Ontario, New Brunswick, Nova Scotia, and Prince Edward Island, have relevant statutes or regulations that govern animal research, teaching, and testing.

Canadian Council of Animal Care

The Canadian Council on Animal Care is 'the national peer review agency responsible for setting and maintaining standards for the care and use of animals used in research, teaching and testing throughout Canada' (CCAC website: About CCAC).

The mandate of the CCAC is 'to act in the interests of the people of Canada to ensure through programs of education, assessment and persuasion that the use of animals, where necessary, for research, teaching and testing employs optimal physical and psychological care according to acceptable scientific standards, and to promote an increased level of knowledge, awareness and sensitivity to relevant ethical principles' (ibid.: Mandate). CCAC programs apply in Canada to all animals used by members, individuals, and employees, agents or owners acting on behalf of organizations or businesses registered or operating in Canada that use animals for research, teaching, regulatory testing, or the manufacture of saleable products. By way of example, CCAC members include, among others, Agriculture and Agri-Food Canada; the Association of Universities and Colleges of Canada; Canada's research-based pharmaceutical companies (R and D); Canadian Association for Laboratory Animal Science; Canadian Institutes of Health Research; the Canadian Veterinary Medical Association; Environment Canada; Fisheries and Oceans Canada; Health Canada; the Medical Research Council; the National Cancer Institute of Canada; the National Research Council; the Natural Sciences and Engineering Research Council (NSERC); and the Canadian Food Inspection Agency.

As noted earlier, in Canada there is a requirement, established by the CCAC, that institutions in which animal-based research, teaching, or testing is conducted establish an Animal Care Committee (ACC). The ACC's terms of reference must reflect and refer to the institution's animal care and use program, including the members of the program and the institution's policies, practices, and procedures.

The CCAC guidelines specify that the relevant ACC must review applications for all animal use (CCAC 1997a). The guidelines are intended to facilitate the work of the ACC by ensuring that a complete and accurate description of the proposed animal use is included in the application. The CCAC guidelines specify that it is the role of the ACC to ensure that approved animal-use protocols support the premises upon which objective review of animal use in science is based:

- that the use of animals in research, teaching, and testing is acceptable only if it promises to contribute to the understanding of environmental principles or issues, fundamental biological principles, or development of knowledge that can reasonably be expected to benefit humans, animals, or the environment;
- that optimal standards for animal health and care result in enhanced credibility and reproducibility of experimental results;
- that acceptance of animal use in science critically depends on maintaining public confidence in the mechanisms and processes used to ensure necessary, humane, and justified animal use; and
- that animals should be used only if the researcher's best efforts to find an alternative have failed. A continuing sharing of knowledge, review of the literature, and adherence to the Russell-Burch 'Three R' tenet of 'Replacement, Reduction and Refinement' are also requisites. Those using animals should employ the most humane methods on the smallest number of appropriate animals required to obtain valid information' (ibid.).

INFORMATION THAT MUST BE INCLUDED IN THE ANIMAL USE APPLICATION

The CCAC requires that all applications for animal use must include the following information (CCAC 1997a):

- *Study summary:* A summary of the primary aims and proposed use of animals in language understandable to a layperson. This should include a description of procedures designed to ensure that animal suffering will be prevented or minimized.
- *Potential benefit of the research:* Applicants must include a statement as to the purpose and potential value of the study as well as evidence of peer review for scientific merit.
- *Replacement alternatives to animal use:* Applicants must consider the use of non-animal models or less sentient animals for the proposed research. The absence of replacement alternatives should be supported by a brief description of the methods and sources used to determine that alternatives were not available, and/or an explanation of the aspects of the protocol that preclude using non-animal models or less sentient animals.
- *Animal model selection:* The specific characteristics of the animal model to be used that make it the most appropriate for the study should be described. Characteristics may include structural, behavioural, physiological, biochemical, or other features or considerations, which may include data from previous studies that make the model most compatible with the stated research objectives. Cost should not be a primary consideration.
- *Reduction in animal use/numbers:* The protocol must include a clear description of the experimental design along with the statistical rationale that supports the size of the control and test groups. The number of animals to be used must be minimized and be scientifically and statistically justified.
- *Refinement of experimental technique:* The ACC and the investigator have a shared responsibility to ensure that the husbandry practices and experimental procedures employed minimize or eliminate physical and/or psychological distress within the limitations imposed by the objectives of the research.
- *Euthanasia:* If applicable, the method of euthanasia, criteria for euthanasia, and the training and competence of the persons performing euthanasia must be described. If surviving animals

are not to be euthanized during the study, or upon completion of the study, a description of what will happen to them is required. Details must also be provided about study endpoints, physical restraint, and invasive or stressful procedures that are to be performed.

- *Hazardous materials:* Appropriate approval for the use of hazardous agents, including biohazards that will be used as part of the study, must be filed with the ACC before commencement of the project. Potential health risks to humans or animals, any special animal care that is required, precautions for personnel, special containment, storage, or waste requirements, animal disposal requirements, and emergency procedures must be described.
- *Transgenic animals:* For animal use protocols that create or use transgenic animals, detail must be provided outlining procedures performed and possible welfare concerns for progeny from transgenic animal creation protocols. A form must accompany the application that details the species, strain, known abnormalities, criteria for monitoring and detecting physical or behavioural abnormalities that are indicative of pain and distress, the need and rationale for biological containment, if any, and the general time frame for production of a novel transgenic strain and reporting back to the ACC on the phenotype obtained.

SPECIAL CONSIDERATIONS FOR ANIMAL USE APPLICATIONS THAT REQUIRE THE USE OF TRANSGENIC ANIMALS – CCAC GUIDELINES ON TRANSGENIC ANIMALS

In 1997, the CCAC published guidelines to assist Animal Care Committee members and investigators to evaluate 'the ethical and technological aspects of the proposed creation, care and use of transgenic animals; to ensure that transgenic animals are used in accordance with the CCAC Statement "Ethics of Animal Investigation" and to ensure that the well-being of Canadians and the environment are protected' (CCAC 1997b).

In this guideline, the CCAC expressly recognizes that transgenic animals provide a powerful tool that researchers can use to develop disease models to better understand human disease. It is hoped that the improved specificity of animal models may logically lead to a reduction of the number of animals used. It also recognizes that there are a series of related ethical concerns which include concerns about animal welfare, human health and environmental harms, animal suffering, and the inadvertent, and ultimately the intentional, modification of the human genome. It is the responsibility of the ACC to ensure that its members are informed about the technological and ethical aspects of transgenic animal use. Specific information that ACCs must have in the application for animal use is documented above under 'Transgenic Animals.'

The obvious gap in the system is that research performed outside of the government or university environments is not directly regulated in Canada. Private firms are not required to be participants in the CCAC program; however, most of them do participate, and most view CCAC compliance as a de facto requirement. Virtually all the large multinational pharmaceutical companies that engage in biomedical research in Canada participate in CCAC programs as a matter of course. Most of the smaller companies have spun off from academic institutions and have evolved within the CCAC system, and see it as a necessary part of doing business. Ideally and logically, comprehensive and binding rules should apply, across the board, to all animal research regardless of where it is performed.

It is anticipated that the CCAC will soon publish new guidelines on genetically engineered animals (Griffin and Gauthier 2007). It is likely that this new guideline will require 'careful welfare assessment of newly generated lines, as much information as possible about genotype, phenotype, and consequences of the modification to the animal' (ibid.: 155). In additional, international harmonization will require the publication of this information into a central repository/database accessible to researchers around the world.

Conclusions

As previously discussed, the use of GM animal models in biomedical research continues to increase. It is viewed as key to unlocking the practical knowledge contained in the human genome sequence and facilitating the development of a new era of genomic medicine. It is at this juncture that the conflicting goals of improving human health and preserving and respecting animal life (including upholding the 3Rs) collide. The use of animal models raises a classic paradox, described by one commentator as follows:

> The need for them [model organisms] will only diminish once most of the fundamental mechanisms of biology have been solved to allow the

> greater use of both human tissue cultures and in silico methods for drug discovery. To reach that point, however, requires the extensive use of model organisms. Given the enormous number of unresolved questions that remain in biology, even if the use of model organisms changes over time, they will remain an integral research tool for molecular biologists. (Hunter 2008: 719)

From a pragmatic standpoint, the appropriate balancing of conflicting social goals is useful to help us clarify the best path forward. The focus group discussions that support this volume provide a wealth of information about how individuals think about the genetic modification of animals; this chapter specifically considers how participants view the use of genetically modified animals in the biomedical arena. The value of the information gleaned from the focus group discussions is maximized when reflected back on the current approaches to the use of transgenic animals in research by companies actively working in the area as well as on current systems of oversight. In light of specific information about what people from relevant groups of interested people think, it is possible to consider whether the key issues they raise are adequately addressed under the current oversight regime and, if not, how the regime might be improved.

As a result of this exercise, the following conclusions can be drawn:

- Biotech companies and the majority of focus group participants accept that animal research involving genetically modified animals is acceptable if the research is intended to benefit human health (for serious medical purposes), animal suffering is minimized, and alternative, non-animal methods are not available.
- There is evidence of substantial commonality in the consideration and conditional acceptance of transgenic animals in research, as described above, by BIO, GlaxoSmithKline, Novartis; this conditional acceptance is paralleled to a large extent by discussions among individuals in the Health Researchers and Scientists focus groups.
- There is evidence of substantial commonality between FRAME and individuals in the Animal Justice focus group in the consideration of research involving transgenic animals. The ultimate goal of stopping animal research through the development of alternatives is widely supported by participants in all focus groups.

- Given the conflicting opinion on the relative merit of transgenic animal model approaches to biomedical research, evidence demonstrating its utility (or non-utility) is needed.
- Effort should be directed towards the development of non-animal alternative methods that will decrease the number of animals used in biomedical research. Researchers, regulators, and funding agencies must recognize that mouse and other animal models are not always good analogues of the human target. To the extent possible, the ability of the funding lever to impact the development of alternative methods should be exploited.
- Several focus group participants expressed concern that the goals of biotech companies and of society are in conflict. The positions articulated by BIO, GlaxoSmithKline, and Novartis on their promotional websites with respect to animal research and animal welfare may support this generalization. With regard to the responsible use of animals in research, there appears to be substantial agreement of principles between focus group participants and (at least officially stated) industry positions.
- There is general agreement between BIO, GlaxoSmithKline, and Novartis and focus group participants that the oversight of transgenic animal research needs to be improved. Ideally, all animal research, regardless of where it is performed, should be formally captured in a coherent system of oversight.
- There is substantial, though far from complete, agreement between industry and focus group participants that the public should know more about how animals are used in research generally and in transgenic animal research in particular. Many focus group participants expressed the view that the public should be actively engaged in consultative processes about animal use in research and in improving existing oversight mechanisms. Despite a marked interest in improving oversight mechanisms, it seems likely that Canadian publics would be generally supportive of the general approach to oversight in this country were existing gaps to be closed and a forward-thinking approach to the collection and analysis of data to ensure that animal research be used only in appropriate circumstances.
- The CCAC's intention to clarify its guidelines on transgenic animals is welcomed. Its intention to mandate better welfare assessments of newly generated transgenic lines and to collect detailed information about genotype, phenotype, and consequences of the

modification is a much-needed first step towards improving the oversight of animal research (Griffin and Gauthier 2007). International harmonization of similar efforts around the world would demand publication of the data generated into a central repository/database accessible to researchers and policymakers around the world.

References

Balls, M., and Fentem, J.H. 1999. The validation and acceptance of alternatives to animal testing. *Toxicology in Vitro* 13: 837–46.

Beck, J.A., Lloyd, S., Hafezparast, M., et al. 2000. Genealogies of mouse inbred strains. *Nature Genetics* 24: 23–5.

Biotechnology Industry Organization (BIO) website. Animals needed for research key points. http://bio.org/foodag/animalpoints.asp?p=yes.

– Statement of ethical principles for the care and use of animals in biotechnology research. http://www.bio.org/bioethics/background/animals.asp.

Brumfiel, G. 2008. Animal-rights activists invade Europe. *Nature* 451: 1034–5.

Canadian Council on Animal Care. About CCAC. http://www.ccac.ca/en/About_CCAC/About_CCAC_Intro.htm.

– CCAC Mandate. http://www.ccac.ca/en/About_CCAC/About_CCAC_Mandate.htm.

– CCAC Members. http://www.ccac.ca/en/About_CCAC/About_CCAC_Members.htm.

– 1993. *Guide to the care and use of experimental animals*. 2nd edition. Ottawa, ON: CCAC.

– 1997a. Guidelines on transgenic animals. http://www.ccac.ca/Documents/Standards/Guidelines/Transgenic_Animals.pdf.

– 1997b. Guidelines on animal use and protocol review. http://www.ccac.ca/Documents/Standards/Guidelines/Protocol_Review.pdf.

– 1997c. *Resource*. Spring/Summer, Supplement.

– 2006. Survey of animal use. http://www.ccac.ca/en/CCAC_Main.htm.

Caulfield, T.A., et al. 2003. Genetic technologies, health care policy and the patent bargain. *Clinical Genetics* (63): 15–18.

Erickson, J. 2006. Translation research and drug development. *Science* 312: 997.

European Partnership for Alternative Approaches to Animal Testing. 2007. Progress report: Where did we really make a difference? http://ec.europa.eu/enterprise/epaa/progrep.htm.

Festing, S., and Wilkinson, R. 2007a. The ethics of animal research. *EMBO Reports* 8: 526–30.

– 2007b. Response by Festing and Wilkinson. *EMBO Reports* 8: 796–7 (letter).

Finegold, D., and Moser, A. 2006. Ethical decision-making in bioscience firms. *Nature Biotechnology* 24: 285–90.

Freyhofer, H.H. 2004. *The Nuremberg medical trial*. New York: Peter Lang Publishing.

Fund for the Replacement of Animals in Medical Research (FRAME). About Frame. http://www.frame.org.uk.

– Transgenic and other GM animals. http://www.frame.org.uk/page.php?pg_id=48.

Gannon, Frank. 2007. Animal rights and human wrongs? *EMBO Reports* 8: 519.

Gauthier, C. 2007. The institutional animal care committee: Keystone of international harmonization. *Alternatives to Animal Testing and Experimentation* 14: 157–61.

GlaxoSmithKline website. The role of transgenic animals in biomedical research. http://www.gsk.com/research/about/about_animals_roles.html.

– 2008. Corporate responsibility report: Access to medicines. http://www.gsk.com/responsibility/cr-review-2007/downloads/access-to-medicines.pdf.

Gottlieb, S., and Wheeler, M.B. 2008. Genetically engineered animals and public health: Compelling benefits for health care, nutrition, the environment and animal welfare. http://www.bio.org/foodag/animals/ge_animal_benefits.pdf.

Griffin, G. 1998. The ethics of animal use for research purposes. *Sepsis* 2: 263–6.

Griffin, G., and Gauthier, C. 2007. Facilitation of an international approach for data sharing and acquisition in relation to genetically-engineered animals. *AATEX* 14: 151–6.

GTC Biotherapeutics website. ATryn – Recombinant Human Antithrombin. http://www.gtc-bio.com/products/atryn.html.

Hackam, D.G., and Redelmeier, D.A. 2006. Translation of research evidence from animals to humans. *JAMA* 296: 1731–2.

Harvard College v. Canada (Commissioner of Patents), [2002] 4 S.C.R. 45, 2002 SCC.

Horrobin, D.F. 2003. Modern biomedical research: An internally self-consistent universe with little contact with medical reality? *Nature Reviews Drug Discovery* 2: 151–4.

Hudson, M. 2007. Why do the numbers of laboratory animal procedures conducted continue to rise? An analysis of the Home Office statistics of scientific procedures on living animals: Great Britain 2005. *ATLA* 35: 177–87.

Huggett, T. 2008. When animal rights turn ugly. *Nature Biotechnology* 26: 603–5.

Hunter, P. 2008. The paradox of model organisms, EMBO Reports 9: 717–20.

International Conference on Harmonization. 1997. ICH Guidance E6. Good clinical practice: Consolidated guideline. http://www.hc-sc.gc.ca/dhp-mps/prodpharma/applic-demande/guide-ld/ich/efficac/e6-eng.php.

International Mouse Strain Resource website. http://www.informatics.jax.org/imsr/IMSRSearchForm.jsp.

Kling, J. 2007. Biotech for your companion? *Nature Biotechnology* 25: 1343–5.

Kmietowicz, Z. 2002. Half as many animals used in medical research as 20 years ago. *BMJ* 324: 134.

Kroeger, M. 2006. How omics technologies can contribute to the '3R' principles by introducing new strategies in animal testing. *Trends in Biotechnology* 24: 343–6.

Lander, E.S., et al. 2001. Initial sequencing and analysis of the human genome. *Nature* 409: 860–921.

Lei, H.H. 2006. ATryn manufactured in genetically engineered goats. *Boston Globe*, 3 June.

Manis, J.P. 2007. Knock out, knock in, knock down: Genetically manipulated mice and the Nobel Prize. *New England Journal of Medicine* 357: 2426–9.

Novartis. Animal welfare policy. http://www.corporatecitizenship.novartis.com/downloads/business-conduct/AWO_Policy_ECN.pdf.

– Transgenic animals. http://www.corporatecitizenship.novartis.com/business-conduct/responsible-rd/animal-welfare/transgenic-animals.shtml.

Nuffield Council on Bioethics. 2005. The ethics of research involving animals. http://www.nuffieldbioethics.org/go/ourwork/animalresearch/publication_178.html.

Pharmaceutical Research and Manufacturers of America (PhRMA). 2007. *Drug discovery and development: Understanding the R and D process*. Washington, DC: PhRMA.

– 2008. *Pharmaceutical industry profile*. Washington, DC: PhRMA.

Pound, P., et al. 2004. Where is the evidence that animal research benefits humans? *BMJ* 328: 514–17.

Pritchard, F.J., et al. 2003. Making better drugs: Decision gates in non-clinical drug development. *Nature Reviews Drug Development* 2: 542–53.

Psaty, B.M., and Charo, R.A. 2007. FDA responds to Institute of Medicine drug safety recommendations – In part. *JAMA* 297: 1917–20.

Rollin, B.E. 2006. The regulation of animal research and the emergence of animal ethics: A conceptual history. *Theoretical Medicine and Bioethics* 27: 285–304.

Russell, W.M.S., and Burch, R.L. 1959. *The principles of humane experimental technique*. London: Methuen.

Sanger Centre. Human genome project shows the wonder and the mystery of humankind. http://www.sanger.ac.uk/HGP/publication2001/mainrelease.shtml.

United Kingdom Home Office. 2006. Statistics of scientific procedures on living animals. http://www.homeoffice.gov.uk/rds/scientific1.html.

United States Food and Drug Administration. 2005. FDA announces important changes and additional warnings for COX-2 selective and non-selective non-steroidal anti-inflammatory drugs (NSAIDS). http://www.fda.gov/CDER/Drug/advisory/COX2.htm.

– 2007. Subcommittee on Science and Technology. FDA science and mission at risk. http://www.fda.gov/ohrms/dockets/ac/07/briefing/2007–4329b_02_00_index.html.

Wall, R.J., and M. Shani. 2008. Are animal models as good as we think? *Theriogenology* 69: 2–9.

Waterston, R.H., et al. 2002. Initial sequencing and comparative analysis of the mouse genome. *Nature* 420: 520–62.

Watts, G. 2007. Alternatives to human experimentation. *BMJ* 334: 182–4.

World Medical Association. 1964. Declaration of Helsinki (as amended). http://www.wma.net/e/policy/b3.htm.

Zambrowicz, B.P., and Sands, A.T. 2003. Knockouts model the 100 best-selling drugs: Will they model the next 100? *Nature Reviews Drug Discovery* 2: 38–51.

Zambrowicz, B.P., Turner, C.A., and Sands, A.T. 2003. Predicting drug efficacy: Knockouts model pipeline drugs of the pharmaceutical industry. *Current Opinion in Pharmacology* 3: 563–70.

7 Human Health Care: The Promise of Animal Biotechnology

NOLA M. RIES

The instrumental use of animals for human benefit is expanding through transgenic technologies, including the technological capacity to modify animals at the genetic level to enhance their suitability to meet human health needs. Numerous possibilities exist to use animal biotechnology to benefit human health. Through transgenesis, goats, cattle, rabbits, and other mammals can produce proteins in their milk to be extracted to develop pharmaceuticals for human use. Pigs can be genetically modified to produce organs and tissues compatible for transplantation in people. Transgenic animals can be engineered for immunity against diseases that afflict them and that can be transmitted across species to cause human disease. Livestock raised for human consumption can be genetically altered to produce more nutritious meat and milk, to eliminate allergenic components, or to reduce the environmental impacts of industrial animal production. Most of these applications are still at the research stage, but some may enter the marketplace or medical practice sooner rather than later.

This chapter discusses these various uses of transgenic animals, all aimed at promoting human health. The chapter takes an expansive view of health, which includes the use of animals for direct therapeutic purposes such as drug production and organ supply, but also their use as sources of nutrition for humans, as disease vectors for humans, and as living beings whose cultivation by humans has environmental consequences. To date, clinical applications of animal biotechnology are few and far between. Fifteen years ago, some biomedical researchers and clinicians expected that, before 2010, organs harvested from transgenic pigs would be in use to address pervasive shortages of human donor organs. Similarly, companies seeking regulatory approval for drugs produced from goat and rabbit milk anticipated that such pharmaceuticals would already be on the market. The slow trickle of human health applications from biomedical research laboratories attracts relatively little public attention, especially when compared to the vigorous public debates in some countries over agricultural biotechnology and the adoption of genetically modified crops. Xenotransplantation stands alone as a biomedical application of animal biotechnology that has been widely debated in public.

In 2008, reports from the United States Food and Drug Administration and the European Food Safety Authority announced that food from certain cloned animals was assessed as safe for human consumption. These findings and ensuing media attention drew public attention to the prospect of using animal biotechnology to produce novel foods, but for many the idea of producing pigs whose meat is rich in heart-healthy omega-3 fatty acids and whose manure is less toxic to the environment, creating chickens resistant to avian influenza, and producing pharmaceuticals from goat milk to treat human diseases still seems the stuff of science fiction.

But, as this chapter discusses, this is the stuff of scientific reality and the stuff that demands public consideration of the extent to which humans ought to genetically alter animals for their own ends. Humans have, of course, manipulated animals for thousands of years, through hunting of wild species (sometimes to extinction), domestication and selective breeding to enhance traits desirable for human purposes, use of animals for sport, entertainment, and scientific experimentation, and, in recent years, massive-scale industrialized production of animals for food. The sheer growth of human populations around the world, our sprawling tendencies, and consumption of natural resources also manipulates animals. In its capacity to alter and manipulate animals, biotechnology is arguably not a novel human activity, but simply a different tool to do what we have always done. Yet the ways in which biotechnology can permit fundamental changes to animals and such novel uses forces us to 're-think our relationship with the natural world.' This quotation is taken from a participant V3 in our focus group composed of health professionals, genetic counsellors, patient advocates, and others who work in the health sector. This chapter incorporates observations and comments from participants in our various focus groups, especially their hopes and concerns about our escalating ability to manipulate animals to serve human health ends.

Fundamental ethical questions emerge from the focus groups' discussions of animal biotechnology. When is it justifiable to use biotechnology to manipulate animals as a means to address human health problems? What other means are available to protect and improve human health and how do those measures rank against the use of animal biotechnology? How do we weigh the benefits and harms, broadly defined, for both humans and animals? How do we ensure fair distribution of the benefits of animal biotechnology between advantaged and disadvantaged individuals and communities? Will animal biotechnology diminish or exacerbate health disparities between the wealthy and the poor? These questions demand careful attention to the values and interests at stake, a core issue explored in all chapters of this book.

Animal Biotechnology for Health

Public opinion research generally reveals greater support for the use of biotechnology for health and environmental purposes and less support for other applications, including agricultural biotechnology and the production of genetically modified crops (Decima 2006). 'Health' can be interpreted narrowly to focus on medical applications of biotechnology that aim to help patients who are already ill and need health care. It can also be defined more broadly to encompass disease prevention and health promotion. Indeed, the Constitution of the World Health Organization defines 'health' as 'a state of complete physical, mental and social well-being and not merely the absence of disease or infirmity' (World Health Organization 1948). In examining uses of animal biotechnology for human health applications, this chapter adopts a broad conception of health. It considers applications of animal biotechnology that may lead to clinical uses for patients with disease and disability, and also uses of animal biotechnology that create conditions that promote human health, including healthier foods and safer environments. Importantly, the focus group participants also adopted a broad perspective on health and emphasized their beliefs about the interconnections between humans, other animals, and the natural environment. As a participant in the Agricultural Producers focus group expressed it: 'There used to be, in my mind, animal health, and then there was human health. And now there's such a link between them that the jump is there now from animal health to human health' (Agricultural Producers stakeholder V6).

The following discussion of animal biotechnology for health considers four contexts: medical treatment, disease prevention, food, and environment. It begins in the medical context and considers two key applications: xenotransplantation and biopharming. Next, the discussion turns to health in the disease-prevention context and considers applications of biotechnology to reduce risks of zoonotic infections that pass from animals to people, as well as interventions that mitigate animal disease burdens. It next considers health in a food context and the use of biotechnology to produce animals for human consumption that provide enhanced nutritional benefits or that have harmful components, such as allergens, removed. This context represents another way in which animal biotechnology can be used to prevent disease, but focuses on healthy nutrition. Finally, the discussion of health and environment focuses specifically on hazards that arise from modern forms of intensive animal farming, where huge amounts of livestock waste can create harmful environmental contaminants. Some of these applications were presented briefly in Mickey Gjerris's overview of animal biotechnology in chapter 2; the following section aims to provide further detail to explain human health applications before turning to a fuller analysis of the values at stake in this context.

Applications of Animal Biotechnology for Health

Medical Treatment Applications

Biomedical applications of transgenic research focus on xenotransplantation, production of pharmaceutical proteins in animals, and the use of animals as models for human disease. Lorraine Sheremeta's chapter on animal biotechnology in research (chapter 6) focuses on the latter application, so this section addresses xenotransplantation and biopharming.

XENOTRANSPLANTATION

According to one estimate, 'approximately 250,000 people are currently only living because of transplantation of an appropriate human organ' (Niemann and Kues 2003: 296). In the United States, approximately 94,000 people were on an organ transplant wait list at the end of 2006 (United States Department of Health and Human Services 2007), and approximately 20 to 30 per cent of these people die while waiting. Some countries, including China, Pakistan, and India, are notorious for illegal trade in human organs, driven by the demand for life-saving transplants by those who can afford to buy an organ on the black market and desperate poverty that impels people to sell organs. In 2007, the *Lancet* medical journal reported that 'in some villages in poor areas of Pakistan, almost no one has both kidneys' (Lancet 2007: 1901). The supply of human organs may be increased through greater public education campaigns encouraging donation, adoption of presumed consent laws that authorize organ removal upon death unless the deceased specifically registered a prior objection, and legalization of commercial sale of organs. Such measures, however, are still unlikely to increase demand for organs to meet all needs. Xenotransplantation – if it ever moves from the laboratory to the clinic – would be one way to address the shortfall.

Mohiuddin (2007: 429) summarizes the trajectory of xenotransplantation over the past two decades: 'Progress in this field from the late eighties to the late nineties had been steady, but shrinking funding, ethical and regulatory issues, threats of transmission of infection, and diminished interest by industry have resulted in a significant decline of enthusiasm in this field. But the recent development of genetically modified pigs that are more compatible with humans has reinstated hope for the success of xenotransplantation of organs.' So transgenic pigs may spell renewed hope for researchers, clinicians, and patients (if not for the donor pigs).

Pigs are already used as a source of heart valves and skin grafts for medical use in humans. Domesticated pigs are considered an ideal source of organs for transplantation to humans for several reasons: (1) pig organs are similar in size and physiological characteristics to human organs; (2) the gestation period is short, and pigs produce large litters; (3) piglets grow and reach maturity quickly; (4) pigs are already a domesticated species and can be raised cost-effectively in hygienic conditions (Prather et al. 2003; Niemann and Kues 2003). Safety concerns with xenotransplantation relate to transmission of animal infections to the human recipient. In pigs, porcine endogenous retroviruses (PERV) pose at least theoretical infection risks for human recipients of pig tissues and organs, and acute or long-term rejection of the transplanted pig tissue or organ is also a concern. Pigs can, however, be genetically modified to produce human proteins to trick the recipient's body into accepting it as a compatible organ. Experimental transplantation of transgenic pig organs into non-human primates (such as baboons) has demonstrated the success of this technique (Bach 1998), though long-term survival after xenotransplantation has yet to be established. At best, baboons who have received pig organs have survived in the range of six months (Mohiuddin 2007).

Xenogenic cell therapy may also be used to treat degenerative diseases (e.g., Parkinson's disease) or severe injury (e.g., spinal cord injury) in humans. Islet cells from pigs have been transplanted to patients with diabetes to help them produce insulin, and neural cells from fetal pigs have been transplanted to patients with Parkinson's disease and Huntington's disease (Niemann and Kues 2003). Xenotransplants of neural, optical, cardiac, and other cells have been conducted between non-human animals (such as transplanting cells from transgenic bovine fetuses to mice or rats) to treat brain, eye, and heart diseases. Advances in human stem-cell research may, however, overtake research in xenogenic cell therapies, particularly if technological advances permit the creation of human stem cells from adults rather than embryonic sources.

In 1996, it was anticipated that organs from transgenic pigs would be in use for human transplantation within five to ten years (Jones 1996). In 2002, however, a *Nature Biotechnology* editorial complained: 'We are still a very long way from ever turning this research into a clinical reality. What's more, the inability of current detection technologies to verify that transplants are free of viral contamination could condemn the field to regulatory oblivion' (Nature Biotechnology 2002: 203). Even though xenotransplantation was – and is – years away from the clinic, numerous jurisdictions have invested resources into public and professional engagement on the subject and the development of position statement and ethics guidelines. A 2004 review of public opinion surveys drawn from 35 sources in 23 countries revealed divided views on xenotransplantation: 'On average 40% found xenotransplantation morally acceptable . . . 54% found the application potentially useful, 57% found it risky, and 42% believed that further research should be encouraged' (Hagelin 2004: 555). The review also concluded that regulatory and scientific stakeholders focus on research aspects of xenotransplantation, while members of the general public are principally concerned with the ethics and safety of animal-to-human transplants.

In Canada, the Canadian Public Health Association (CPHA) carried out public consultation on xenotransplantation and issued a final report in 2001. Interestingly, it found that 'when generally uninformed Canadians were asked if Canada should proceed with xenotransplantation, the majority said yes. However, as they became more informed, a shift occurred, and the majority of informed Canadians said no, Canada should not proceed' (CPHA 2001: 10). Yet, a correlation between increased knowledge and a shift to negative views on xenotransplantation is perhaps not surprising. The fact of organ shortages is relatively common knowledge, especially with public advertising campaigns encouraging people to register as organ donors. The popular media has publicized stories of illegal trade in human organs, an odious consequence of the shortage of voluntarily donated organs. When presented with the idea of xenotransplantation, those who overcome an initial 'yuk' reaction may be inclined to the view that cross-species transplants are acceptable provided they are safe for the human recipient. After all, if pig organs can reduce harms of organ shortages and illegal trafficking, xenotransplantation is beneficial. Yet, this assessment of the potential benefits of xenotransplantation for humans does not take account of other interests at stake, including animal welfare concerns, worries about unanticipated harms, and objections to high-tech solutions where other, less troubling measures can be used to address the health issue. These concerns are explored in further detail in the 'Values at Stake' section, below.

ANIMALS TO PRODUCE PHARMACEUTICALS

Animals can be used as 'bioreactors' to produce proteins in their milk or other bodily fluids to be extracted and processed into pharmaceuticals to treat human disease. Drugs to treat alpha-AT deficiency, cystic fibrosis, coronary clots, heparin resistance, and Pompe's disease have been produced variously in the milk of sheep, goats, and rabbits and used in phase II and III clinical trials. According to one scientist, this 'biopharming' mode of therapeutic protein production 'offers a safe and renewable source of clinically important proteins that cannot be produced as efficiently in adequate quantities by other methods' (Rudolph 1999: 367). In 2003, it was anticipated that these products would be on the market by 2006 (Niemann and Kues 2003), but to date only one biopharmed drug – ATryn® – has received regulatory approval. ATryn®, produced by GTC Biotherapeutics Inc., is approved by the European Medicines Agency and the United States Food and Drug Administration for use to prevent thromboembolism (blood clots and arterial blockages) in high-risk surgical patients with a congenital deficiency in normal blood-clotting proteins.

The Dutch company Pharming Group is conducting clinical trials with its drug Rhucin®, which is produced in the milk of transgenic rabbits and used to treat a rare genetic disorder that causes painful and sometimes fatal episodes of soft-tissue swelling (angioedema). Pharming is also producing human lactoferrin in the milk of transgenic cows. This immune-stimulating protein is found naturally in the human body, but lactoferrin supplementation may boost resistance to various infections. Lactoferrin derived from non-transgenic cattle is currently used in Japan as a nutritional supplement and food additive. With promise in the Japanese market, Pharming has obtained a patent in Japan on its recombinant human lactoferrin, with further uses in food and sport nutrition products.[1]

Some preliminary research has attempted to gauge public attitudes towards biopharming. One study (Duguay, Katsanis, and Thakor 2003) examined the influence of five factors on the consumer acceptability of transgenic biopharmaceuticals: (1) consumer knowledge of the technology; (2) perceived risks of the technology; (3) relative advantage of the technology; (4) concordance with personal values; and (5) perceived complexity of the science. The belief that drugs made in transgenic animals would benefit patients promoted consumer acceptance. In contrast, perceptions that the technology is too complicated and concerns about 'the generation of new types of animals, custom-made for every application' (ibid.: 85) undermined acceptance.

Our focus groups were asked for their reactions to an example of an insulin-producing cow. The prevalence of diabetes and the potential for significant human benefit were cited as factors in favour of this biopharming example. A participant in the Alternative Agriculture focus group would only accept this application if 'control measures are in place to make sure that those cows don't end up . . . in my milk supply' (V1). Interestingly, a participant in the Health Researchers focus group rejected this animal biopharming example and favoured plant-based biopharming instead, since production of insulin from a cow requires expensive milk storage facilities. For this person, biopharming to produce drugs is clearly acceptable, but it should be done in a way that is most cost-effective. Only one participant was uneasy about an insulin-producing cow on the grounds that it involves the 'introduction of something foreign to [animals, that is] not for their benefit' (Health Care Providers and Patient Advocates stakeholder V5). In this example, the potential benefit of using animal biotechnology to treat a well-known and widespread disease clearly outweighed concerns about tinkering with the natural essence of the cow.

Zoonotic Disease Reduction

Biotechnology offers means to reduce the burden of disease that afflict animals, as well as zoonotic infections that can spread from animal to human populations. Our focus groups were not asked to consider a use of animal biotechnology to reduce zoonotic disease, but examples include the use of transgenic technology to develop animals resistant to common infections, such as mastitis (a bacterial infection of the mammary gland) in dairy cattle, and prion disease, such as bovine spongiform encephalopathy (BSE or 'mad cow' disease). Disease outbreaks in agricultural animals have resulted in widespread culls, such as a highly publicized slaughter of 10 million sheep and cattle in 2001 in the United Kingdom to counter a foot and mouth disease outbreak. In Canada, approximately 15 million ducks, geese, and chickens in commercial poultry operations were killed to contain a 2004 outbreak of avian influenza (Canadian Poultry Industry Forum 2004: 6).

Zoonotic diseases are a leading cause of human illness. These include bacterial infections (e.g., salmonella and *E. coli*), parasites, viruses (e.g., rabies, avian influenza), and other unconventional or emerging diseases such as variant Creutzfeldt-Jakob disease. Researchers have been working on the development of genetically modified animals that are incapable of spreading specific diseases to humans or that can disrupt natural populations to mitigate disease spread. A UK company, Oxitec,[2] has created a genetically modified mosquito that produces non-viable offspring. When the laboratory-produced male mosquitoes are released into a target area, they compete with wild males to mate with females, and the offspring of the GM males do not survive past the larval stage. Releasing GM males in sufficient numbers would, it is hoped, allow them to outnumber and out-compete with wild males and decimate the subsequent generation of mosquitoes, though concerns persist about long-term ecosystem disruption (Pew Initiative on Food and Biotechnology 2004). In April 2008, a Malaysian medical research institute announced plans to release Oxitec's GM mosquito as a method to fight mosquito-borne dengue fever. Southeast Asia is one of the areas most seriously affected by this disease worldwide; in 2005 (the most recent year for which the World Health Organization posts data), Malaysia had nearly 17,000 cases of dengue fever and dengue hemorrhagic fever.[3] Malaysian environmental advocates, however, question the long-term ecological impacts of introducing the GM mosquito (Cyranoski 2008).

In Australia, researchers with the country's national science agency, the Commonwealth Scientific and Industrial Research Organisation, are experimenting with GM technology to produce chickens resistant to infection with avian influenza, especially the highly virulent H5N1 strain. H5N1 avian flu has caused wide-scale outbreaks of disease within the past several years. In a 2004 outbreak in Asia, over 100 million chickens died or were slaughtered to contain spread of the disease (World Health Organization 2004a). By the start of 2007, H5N1 outbreaks had been reported in bird species (domestic and wild) over 40 countries, massive poultry culls destroyed over 200 million birds in Asia, and some countries imposed severe restrictions on poultry production, including bans on keeping flocks in backyards. Humans in close contact with affected birds may become infected, and in rare cases, an infected human can pass the disease to another person. As of April 2009, 417 human cases of H5N1 infection and 257 deaths had been reported to WHO (World Health Organization 2009). Through genetic modification, chickens can be produced with a heightened ability to fight off infection with H5N1.

While these efforts to reduce zoonotic diseases sound laudable, particularly to address a prevalent disease like malaria in developing countries, the question can be asked in some cases whether alternatives to biotechnology would better address the root of the problem. In the 'mad cow' example, the human practice of feeding ruminant protein to other ruminants led to the emergence of BSE. Those who have faith in regulatory systems may argue that resources should be allocated to stringent enforcement of feed bans rather than genetically engineering cattle to resist BSE. Faith in regulators is constantly tested, however, with periodic scandals and crises, such as the 2008 outbreak of listeriosis in the Canadian food supply and the Walkerton water contamination, discussed below.

Food

Access to a safe and adequate food supply is essential to human health. Diet-related disease is a rapidly growing health problem in developed and developing countries around the world (World Health

Organization 2004b). Higher rates of obesity, type 2 diabetes, cardiovascular disease, and some cancers are linked with modern dietary patterns marked by over-consumption of calorie-dense and nutrient-poor foods. The global health problems of obesity and related conditions are now more common than hunger and starvation. The issue of food insecurity, however, encompasses both problems: people are considered food insecure when they have inadequate access to sufficient calories and also when they obtain excessive calories from foods high in fat, sugar, and salt and low in nutritional value. Animal biotechnology to enhance the health benefits of food are one means to address food insecurity. Plant and animal sources of food can be genetically modified to reach maturity more quickly, to thrive in a wider range of production conditions, to be devoid, or have reduced amounts, of unhealthy or harmful components (such as saturated fats or allergens), and to deliver beneficial nutrients.

Food products with added bioactive components are already available in the marketplace. Consumers may purchase yogurt with added beneficial bacteria to improve digestion, eggs with higher levels of omega-3 fatty acids to improve cardiovascular health, and cereals with added fibre to promote healthy bowel function. In some cases, these products simply involve adding an ingredient during processing, such as adding fibre to processed breakfast cereals. In other cases, modification occurs earlier in the food production cycle, such as feeding flaxseed to chickens to boost omega-3 content in their eggs. Extending these efforts even further, transgenic technology may be used to enhance the nutritional content of animal milk and meat consumed by humans. Transgenic goats and cattle have been developed in research settings to express specific human proteins in their milk that, if consumed by humans, could have beneficial health impacts, such as improved immune response against pathogens. Genetic modification could also produce cows that provide lower-fat milk. Genetic modification of animals may also allow for the elimination of certain food bioactives that trigger adverse human reactions. GM dairy cattle can produce lactose-free milk or allergens could be removed from foods.

Biotechnological advances already seek to enhance the nutritional profile of food plants and develop functional foods, foods that are consumed for a health benefit beyond meeting basic nutritional needs. In addition to meeting nutrition needs, companies could generate new markets and financial rewards if health-conscious consumers are willing to pay more for these products. One commentator notes: 'Designer milks, specialty milks, or humanized milks may be competing in the next ten years to capture part of the global dairy product market worth $400 billion annually' (Karatzas 2003: 138). The growing range of food products with added ingredients that make health claims are increasingly attracting consumers, and routine exposure to such products may encourage consumer acceptance of food with added health benefits that are the result of animal biotechnology. Some of our focus group participants, however, were sceptical about the example of an omega-3 pig, considering it to be a health fad that could have unanticipated adverse health consequences and not worth pursuing since other dietary sources of the nutrient are available. Again, this raises the question of whether animal biotechnology is really necessary for human health benefits, or whether more conventional, safer, and ethically acceptable approaches could be used.

Environment

The connection among livestock production, environmental contamination, and adverse human health impacts is demonstrated by the tragedy that occurred in the town of Walkerton, Ontario, in 2000. Municipal water supplies were contaminated with *E. coli* bacteria that originated from cattle manure spread on the fields of a neighbouring farm. Over 2300 people became sick after drinking tap water, and seven people died; others have had lingering health effects (Walkerton Inquiry Report 2002).

The farm involved in the Walkerton contamination was a relatively small operation, and the farmer was not faulted in any way for his practices; rather, failures in water-safety inspection and regulation were to blame. Intensive livestock operations – involving hundreds or thousands of animals – pose even further risks of environmental contamination than small farms if wastes are not managed appropriately. One hog, for example, is estimated to produce one tonne of manure every year (Auditor General of Canada 2005), and in countries like Canada, more intensive hog farming is now the norm. The ten-year period from 1991 to 2001 saw the average number of hogs per farm increase from 345 to over 900, with some mega-farms having up to 10,000 hogs (ibid.). Producing 1000 hogs translates into producing 1000 tonnes of manure annually.

The environmental burden of livestock waste could be mitigated through use of animal biotechnology to produce less polluting animals. The Enviropig^{TM}, for instance, is genetically modified to produce waste with reduced levels of phosphorous (Golovan 2001), an element that can be a serious source of contamination. Our focus groups pondered the acceptability of a low-phosphorous pig. A few participants expressed resigned acceptance of this application: 'There are so many pigs in production as it is, I figured, "Well, if there are going to be that many pigs anyway, we might as well . . . help nature along a bit by having them produce less phosphorous [especially] given the environmental crisis we have"' (Health Care Provider and Patient Advocate stakeholder V1). Other participants worried about unforeseen future consequences, to animal, human, and environmental health: 'First of all [long-term effects] in the people eating the pigs and then . . . you're maybe reducing the phosphorous into the environment but what other ecological effects are there? If it allows for larger pig factories then, you know, there's other by-products . . . from that' (Alternative Agriculture stakeholder V1).

Each of the applications of animal biotechnology discussed in the preceding section raises potential benefits for human health, some perhaps more compellingly than others. These benefits must be weighed against competing concerns, and this chapter now turns to an analysis of how stakeholders balance benefits and harms within specific value frameworks.

Values at Stake in Animal Biotechnology for Human Health

Assuming the direct human risks of health-related animal biotechnology uses are reduced to an acceptable level (for example, such that xenotransplantation does not carry any risk above the risk a recipient would face with a human organ), animal biotechnology poses other potential harms. For any human health gains we achieve through genetically altering animals for our purposes, do we lose something of our relationship with animals and take another step down a slope that becomes more slippery with each new manipulation of them? The focus groups articulated these questions throughout their discussion and emphasized the need to evaluate the ethics of human health applications of technology, including questions of justice in resource allocation, treatment of animals, and long-term consequences of human interference with other species. As one focus group member put it, the ability to control animals through biotechnology may lead to 'a loss of respect for other life forms, and maybe a loss of spirituality' (Health Care Provider and Patient Advocate stakeholder V5). This is the flip side of the aspiration that animal biotechnology may promote a deeper understanding of the connections among all living beings, the need to 're-think our relationship with the natural environment' referred to at the outset of this chapter.

Who Benefits?

A dominant question in all applications of animal biotechnology is, Who stands to benefit most? Not surprisingly, animal biotechnology for health is promoted principally for its promise in treating and preventing disease, though some critics counter that the primary beneficiaries are 'the companies that own the technology. Perhaps it is the goal of agribusiness to control our food supply through the introduction of patented genetically manipulated and cloned agricultural products' (Schubert 2007: 283). On this sceptical view, claims of health and environmental benefits for humans mask, at best, a misguided motive and, at worst, a blinkered one concerned dominantly with profit. The Health Care Providers and Patient Advocates focus group commented on profit motives and financial investments at stake in animal biotechnology, especially for companies that may be invested in research to find a blockbuster drug to treat 'heart disease, Alzheimer's, [or] Parkinson's' (V1). The prospect of substantial profits creates an incentive to cover up adverse events that might surface or other 'manipulation[s] of the honesty of science' (V3). One participant, V5, suggested, 'So if, for example, a recessive disease came up in an animal population, there would be strong financial incentives not to let the word out, or your stocks would plummet.'

Interestingly, in discussing who benefits from animal biotechnology – or who *should* benefit – the Health Care Providers and Patient Advocates group focused first on animals that are the subject of research, genetic modification, and commercial production. Although one might expect these participants to be predominantly allied with patients, they expressed a strong view that modification of animals for health-related goals should, ideally, benefit the animals; for example, through advancing knowledge about disease that affects animals, helping create better environmental conditions for animals, or by preserving species at risk of extinction. Evincing a similar position, a participant from the Scientists focus group insisted that 'we should stand up and tell society . . . we care about animals as well. We are not only using the animals' (Scientist V1). However, a participant in the Health Care Providers and Patient Advocates group opined that humans are the primary beneficiary of animal biotechnology, and 'animals benefit as a secondary result of the benefit that humans realize' (V2). Another participant in this group ultimately described the hope for benefits to animals as a 'faint' or 'timid' aspiration (V2), with an acknowledgment (resignation even) that humans will benefit over animals and, among humans, that commercial players in the biotechnology industry will benefit the most.

The concern with benefits to animals ultimately seemed overshadowed by concern with benefits for humans. Applications that stood to have widespread benefits, such as an insulin-producing cow to help treat diabetic patients, tended to have greater support. At times, distinctly personal interests influenced participants' views; animal biotechnology tends to become more acceptable if a close family member with diabetes or in need of a donor organ could benefit. Yet, putting aside familial concerns, participants were concerned that animal biotechnology could exacerbate health inequities.

Economics and Exacerbating Disparities in Health Status and Access to Health Care

As with other forms of technology in health care, the financial cost of animal biotechnology is high, especially at the early stages of development. Costs of producing genetically modified functional food products are also staggering, so much so that companies may abandon promising applications because the expense of getting a novel food product to market is very high and public acceptance of GM foods can evaporate quickly at any report of potential hazards. According to one US biotechnology company, it takes '8–10 years and [a] $50–100 million investment' (Powell 2007: 528) to produce a commercializable GM crop product, an expense that often cannot be recouped unless consumers are willing to pay a premium for the health benefits of the novel item.

In xenotransplantation, the cost of using pig organs and tissues is estimated to be at least as high as with transplants from human sources (Canadian Public Health Association 2001). Canadian data reports the cost of a kidney transplant at approximately $20,000 and a heart transplant at $80,000, with additional expenses for anti-rejection drugs. Xenotransplantation presents additional costs associated with breeding and caring for animals, harvesting organs, ensuring safe disposal of animal remains, long-term follow-up of organ recipients, and management of any future adverse events associated with the xenotransplant.

Members of the Health Care Providers and Patient Advocates focus group had particular concern that animal biotechnology may exacerbate existing disparities in health status and access to health care services, both within and between countries. As one participant observed: 'Beneficiaries are going to be those who are in Western medicine, those who exist in large urban centres, those who have easy access to technology' (Health Care Provider and Patient Advocate V3). This participant also noted the problem of making medical advances from industrialized countries available for less developed regions: 'At the global level . . . it will be the Third World countries that won't benefit. It will widen that inequity. But I think it's also going to widen local inequities.'

Biotechnology may well exacerbate the so-called 90/10 problem, where a mere 10 per cent of biomedical research funds are allocated to address problems that account for 90 per cent of the global disease burden (Resnik 2004).

Animal Welfare Concerns

Many examples discussed in this chapter involve domesticated animals that humans currently use for agricultural purposes, such as pigs, goats, sheep, cattle, and chickens. Does the fact that people already use these animals for food mean they are more likely to accept biotechnology applications that alter these animals to produce healthier sources of food? In contrast, is the genetic alteration of animals for their use in entirely new ways, such as xenotransplantation, or as bioreactors to produce pharmaceuticals, less acceptable? The novelty of the use, in itself, does not seem to determine the response. For some stakeholders, concerns about animal welfare are dominant considerations. The extent to which they value animal life and believe that a certain type of life is 'normal' for particular animal species strongly influences their attitudes towards biotechnology.

Our Health Care Providers and Patient Advocates focus group worried that if experimental animals are raised for potential health care applications, many animals will be killed until research establishes the experiments' safety. One participant recounted a situation where pigs were raised in sterile conditions but were 'killed generation after generation. They wouldn't even put them into the food chain. They were ready to transplant their organs, but wouldn't put them into the food chain' (Health Care Provider and Patient Advocate V3).

Participants expressed grave concern for loss of animal life with no benefit to humans (neither medical benefit nor benefit as a food source); some even expressed concern about sacrificing or harming animals where human benefit was likely, such as toxicity testing on animals to prevent harmful products from entering the human marketplace. A participant in the Animal Justice focus group stated: 'For me, the perspective is . . . caring about morality across the board, and not causing harm [to animals] in an effort to prevent other harm [to humans]' (Animal Justice stakeholder V2).

The 'wastage of animal life' (V1) was identified as a principal concern among participants in the Animal Justice focus group. A critic of animal biotechnology cites the huge loss of animal life involved in experiments to create transgenic pigs that produce omega-3 in their tissue:

> [The scientists] began with 1,633 embryos, which they transferred into 14 gilts, which generated five pregnancies that went to term, producing only ten live offspring (two stillbirths), three of which had congenital heart defects and had to be euthanized. So from the 1,633 embryos, the researchers achieved seven (apparently) healthy animals, or an efficiency of 0.4%. In this context, I do not believe it is plausible to speak of animal welfare 'benefits' of the omega-3 pig research. (Fiester 2007: 506)

While these numbers may be inconsequential for those who believe animal embryos do not have interests in being born, killing living animals in the quest for new knowledge troubled many in our focus group. A participant from the Scientists focus group, for instance, acknowledged these poor 'success rates,' noting 'how many pigs had to be killed to do certain risk assessments' (Scientist V6).

Moving animal biotechnology from research to commercial application will, in some cases, require the production of large numbers of genetically altered animals. Similarly, the production, use, and death of animals as a source of organ and tissue transplants for humans raises important ethical issues, while genetic modification of animals to improve their suitability as transplant sources may heighten concerns about human manipulation of animals. The position paper of the Ethics Committee of the International Xenotransplantation Association addresses animal ethics rather briefly, acknowledging that 'the rights of the xenograft source animals are a consideration that has generated controversy' (Sykes et al. 2003: 199). The committee states that ethical concerns are more vexing in regard to the use of primates, such as baboons, rather than other animals, such as pigs, that are already domesticated and routinely used as a food source. It refers to special concerns of some religious groups, vegetarians, and animal rights advocates, but asserts that opposition to human exploitation of animals for human needs 'is not a mainstream view in societies in which meat is eaten, leather goods used, etc.' (ibid.). They further state that 'genetic modifications of xenogeneic [different species] source animals are considered to be acceptable as long as they do not change the overall character of the animal species' (ibid.).

This reference to an animal's character raises the issue of the conditions under which animals are raised for human biomedical purposes. Some commentators argue that production conditions necessary to supply safe tissues or products deprive animals of opportunities to express normal or instinctual behaviours. Fiester, for example, writes:

> Pigs, it turns out, are highly social animals, extremely intelligent, with a curiosity that, unfulfilled, turns into self-destructive or aggressive behaviour . . . In this xenotransplantation project, the alteration of the pigs' environment begins at birth: the pigs are delivered from the sow inside the uterus via cesarean section and placed in a sterile incubator. They are not allowed to suckle; in fact, they have no contact with their mothers at all, and the mother is euthanized after the birth. In their sterile containers, there are no objects to satisfy the pigs' natural need for rooting or intellectual stimulation. They are kept confined, often alone. (2007: 19)

A participant in our Animal Justice focus group echoed these concerns and argued, somewhat rhetorically, that research to benefit humans should be carried out using human subjects. 'We don't do it on humans because it's terrible,' V2 stated. 'So they're deprived of their liberty, they are hurt, they are made to die, they have no normal life with family or social relations.' If this abrogation of a species-normal life is not acceptable for humans, then why is it acceptable for animals? This comment reveals the force with which personal values shape attitudes towards animal biotechnology. On this view, depriving animals of liberty is wrong, whether for biomedical research, food production, or other ways that make animals instruments of human wants.

Altering the Essence of the Species

At what point does genetic alteration change the fundamental character of a species? Is there a line at which alteration becomes too extreme? To take a plant example, a 2008 *Lancet* article on peanut allergies noted that genetic modification of peanuts may attenuate the proteins that cause an allergic reaction, but 'the process of altering enough of the peanut allergens to make a modified peanut that is less likely to cause an allergic reaction would probably render the new peanut no longer a peanut' (Burks 2008: 1543). The author refers here to alterations of the biological properties of the peanut that would make it genetically unrecognizable as a peanut to another scientist who analyses proteins of the plant without knowing the source of the genetic material. In other words, the allergen-free nut might visually look like a peanut, but its biochemistry is very different.

Beyond the biochemistry, genetic alterations (especially transgenic applications where genes are transferred across species) raise aesthetic or metaphysical questions about when a plant or animal becomes something new or different and possibly ethically unacceptable as a result. A participant in our Scientists focus group put it well:

> We debate a lot in transgenics, or there has been a debate in the public as well if you take a human gene and you put it into a pig, do you make a pig a human. If you take a jellyfish, use pig protein . . . do you make a pig a jellyfish? And obviously the answer is no. If you took a human liver and you transplanted it into a pig, would you make a pig a human? No, you wouldn't, but you would give it a metabolic profile that was very similar to a human being, rather than pigs. But if you took a [human] brain and put it into an animal, I think you've really got a question of what that animal is, and I find that unacceptable. (Scientist V2)

The genetic alteration of animals heightens concerns compared to modification of other organisms such as plants. One Health Care Providers and Patient Advocates participant referred to 'the closeness of the human experience to animals' and contrasted it with human experiences or relationships with plants: 'I don't understand plants' nervous system and their ability to observe their environment, and suffer. I don't understand that enough to be empathic with them – to have the same level of empathy with plants that I do with animals' (Health Care Provider and Patient Advocate V5). This tendency to rank other species according to degree of 'humanness' means that some people would likely not be troubled by using biotechnology to create a 'freenut' (as one might call a peanut free of harmful allergens).

A ranking of animals often occurs, too, according to perceptions of the degree to which animals experience suffering, are intelligent, have their own social relationships within their species – in other words, the extent to which people perceive animals as being *like* humans. This fact of relating to certain species more than others might lead some to accept genetic modification of mosquitoes to combat malaria or dengue fever, but oppose raising GM pigs in environments that deprive the animal of opportunities to act on its natural desires and instincts. Ecological concerns may, however, oppose release of GM mosquitoes into the open environment while accepting the production of GM pigs in controlled facilities.

Reiterating the scientist's concern expressed above, a Health Care Provider and Patient Advocate participant (V2) believed it would be 'abhorrent' to grow human brain tissue in a primate's brain for transplantation. An intractable debate in biotechnology concerns perceptions of what is 'natural' and 'unnatural.' As a participant in the Alternative Agriculture group commented: 'It's unnatural processes that are happening and . . . it's not so much that I am against progress . . . but it's how we are getting there' (Alternative Agriculture stakeholder V2).

This debate is reflected in many ways in the issue of genetically modifying animals for health applications. For example, some would not support using cloning technology to revive an extinct species 'because some species are supposed to die out. That's the way it works' (Health Care Provider and Patient Advocate stakeholder V1). Yet, human activity may be seen as having such a pervasive impact on the environment that everything we do impacts nature. This observation suggests that human interference with the environment has, itself, become unnatural and uses of equally unnatural novel technologies may be the only means to address contemporary human problems.

Views about other applications of technology influence perspectives on animal biotechnology more specifically. For example, while focus group participants acknowledged that consumption of salmon is healthy for humans, they had concerns about genetically modifying salmon to grow more quickly and efficiently under aquaculture conditions. Disapproval of aquaculture more generally, largely because of perceived harms to wild salmon, promoted negative views on using genetic modification to further advance aquaculture, even if the resulting food provides health benefits for humans.

What Is the Real Root of the Problem and Is Animal Biotechnology the Answer?

The ability to manipulate animals for our own ends raises a fundamental question: Do we *need* to do this? Does technological capability drive a myopic or distorted vision of solutions to human health problems? Modern health care systems in resource-rich countries emphasize technological fixes, and disease treatment with pharmaceuticals, surgery, and other medical interventions overshadows prevention and health promotion activities. It has been argued that 'medicine, probably more than any other modern practice, needs moral transcendence over technological determinism' (Sorenson 1990: 82).

The issue of whether we *need* to pursue animal biotechnology is raised in all the applications discussed in this chapter. In each case, it is arguable that, indeed, we need to use animals: to provide a solution to an intractable shortage of organs for people who will die on a transplant waiting list; to produce pharmaceuticals; to make food healthier; and to protect humans and other animals from preventable diseases. Yet, probing beneath the surface of each of these claims raises troubling questions.

One can dispute the claim that humans need biotechnology and genetically altered animals to address contemporary health and environmental problems. Recourse to technology is arguably a superficial response to problems with deep roots in modern societies focused on over-consumption (of food and material goods) and industrial production (in agriculture and other sectors) to feed those desires and habits.

In discussing the modification of domesticated food animals, such as a less polluting pig, some focus group participants queried the fundamental source of the problem of excessive pig manure that creates environmental contamination. They identified human desire for pork as the problem. One participant in the Health Researchers group contended that genetic modification here provides 'no value to the pig, and it [is] just an excuse for the human race to keep polluting the land' (V5). Another, in the Health Care Providers and Patient Advocates group, suggested: 'To me, we should be eating hemp if we need more protein, and then that leaves pigs to just live more naturally' (V2). A participant in our Regulators focus group argued we should 'minimize our use of animals and use them where there are no other alternatives' (V6). In many cases, then, the novel biotechnology option may not be the preferred one to address a health problem.

Yet the boundaries of one's academic or professional discipline play an important role in shaping perceptions of problems and the tools used to attempt to solve those problems. In the words of an Alternative Agriculture focus group participant:

> If you have a geneticist who is . . . trying to develop the stress-free chicken, he doesn't see it as his job to find alternatives. He's trying to solve this problem using his discipline which is genetics, and he may not even have a conversation with an animal behaviour researcher who has the alternative suggestions for how to redesign the animal's environment to solve the problems in a different way . . . Each researcher may even have a completely different concept of what progress means for them, and they would be living in a pretty isolated world.' (Alternative Agriculture stakeholder V4)

This statement underscores the need for broad consideration of tools that can be used to address human health problems. Groups with entrenched power and resources are, however, likely to dominate agendas, and simple reallocation of resources from animal biotechnology research to health promotion campaigns does not happen. The 'Eat Hemp, Not GM Pork' movement will gain little traction in this sociopolitical context.

Controversy over high-tech approaches to health concerns is especially evident in the food sector. Industry and academic researchers are increasingly trying to develop 'healthier,' 'value-added' food products, but critics pan the emphasis on using nutrition science and biotechnology to improve nutrition. Pollan (2008) expresses deep scepticism about diets based on the products of food science – processed 'food-like' substances with ingredients added or subtracted to permit manufacturers to make claims about the health benefits of consuming their product. Pollan advocates a simple rule to guide dietary choice: don't eat anything your grandmother would not recognize as food. This rule may apply quite handily to a package of orange and white tubes bearing the name 'cheese strings' or a box of brightly coloured rings labelled 'fruit loops.' Food applications of animal biotechnology are not so obvious, however. A strip of bacon from a transgenic pig that expresses high levels of omega-3 fatty acids would look just like a strip of bacon your grandmother fried with eggs for grandpa's breakfast. While discussion of food labelling is outside the scope of this chapter (see chapter 10 for such a discussion), the non-obviousness of genetically modified foods motivates those who advocate for GM food labels.

Concern with Long-term Consequences

Some focus group participants expressed concern about wilful or unthinking blindness to future consequences of manipulating animals in the search for human health gains. They worried that using animal biotechnology for health applications might lead the way to uses they considered morally repugnant, such as bioterrorism applications, creation of designer humans, and human cloning. They feared increased monoculture and loss of genetic diversity among animal species as well as triggering novel diseases for animals and humans. A participant in the Regulators focus group expressed concern that 'unintentionally introducing detrimental traits . . . could have effects on the animal's long-term survival or productivity or on the environment, cumulative things that are very different – indirect effects that are almost impossible to model or predict over a long time' (Regulator V2).

Governance frameworks, like research ethics approval processes, were considered by one health care provider as too myopic in regard to long-term consequences, as they are perceived to focus on 'utilitarian bottom-line issues' (V4), namely, the immediate gains that may be derived from research. However, many focus group participants agreed that the risk of unanticipated and unforeseen harms ought not to halt animal biotechnology but rather creates an obligation for assiduous monitoring: 'We need to be mindful of the fact that we need follow-up on what's going on, and to watch carefully for those unexpected circumstances to arise so we can deal with them and not just sweep them under the carpet' (Health Care Provider V1). A participant in the Regulators focus group confirmed that there are 'always uncertainties, and we compensate, as regulators, for uncertainties by putting in place risk management measures' (Regulator V1).

Regulating Animal Biotechnology for Health

The final chapter of this book grapples in detail with regulatory challenges posed by animal biotechnology, but regulation issues are raised here since a unique discussion arose in the Health Care Providers and Patient Advocates focus group where a participant spontaneously described a set of questions that should be used to assess the acceptability of an animal biotechnology application. Participants in that focus group acknowledged that some reactions to animal biotechnology are visceral and that it would be preferable to have an analytical and ethical framework to guide decision making about the acceptability or rejection of various applications of animal biotechnology. The following questions were proposed as a framework for assessing the acceptability of the genetic modification of an animal:

1. What is the purpose of the genetic modification? Is the prospect of a beneficial outcome clear or ambiguous? The uncertainty of benefit would result in a lower ranking of acceptability. Is the benefit clear but 'gratuitous' or frivolous? One Health Care Providers and Patient Advocates focus group participant (V2) described the benchmark for a gratuitous activity as something '12-year-olds would do on a Saturday afternoon in the garage.' A clear but nefarious purpose would rank lowest (for example, genetically modifying animals for violent purposes).
2. Who is the beneficiary? The acceptability of a biotechnology application would be higher in situations where animals are a principal beneficiary. If humans are the beneficiary, is the benefit for human health (higher ranking), and is the health concern one that is prevalent in the population (e.g., diabetes) such that the application of animal biotechnology could be quite beneficial. As one participant in our Scientists focus groups stated, the goal should be 'the greatest benefit spread as widely as possible' (Scientist V5).
3. What is the nature of the risk? This question confronts the issue of what we do and do not know. Is the genetic modification 'an enhancement of something that already exists [or is it] the introduction of something unnatural?' To what degree are impacts unknown or controllable?
4. What are the costs, broadly defined? In particular, what is the cost to the animal (for example, degree of suffering, waste of animals in experiments)?

One's values will influence how one interprets and answers questions at each stage of this proposed framework. Deciding which applications are beneficial and to whom, and whether risks and costs are acceptable or not, will depend on one's perspective on all the issues discussed earlier in this chapter.

The Health Care Providers and Patient Advocates focus group underscored the need for informed public input into regulatory decisions. A participant in the Regulators focus group also expressed a desire for more balanced attention to what animal biotechnology can produce: '[If] somebody genetically engineers a cat to have pink fur . . . it will be on every newspaper around the world, but . . . if the cost of development of a new drug was a hundred million dollars less [due to biotechnology] we wouldn't hear' (Regulator V4).

Some participants, however, expressed 'ennui' with endless government studies of issues, when 'nobody seems to want to have the intestinal fortitude to actually take it and run with it and make some decisions about it' (Health Care Providers and Patient Advocates stakeholder V1). There is a desire for transparent regulatory processes that take account of the range of potential benefits and harms of animal biotechnology, but also a desire that a responsible authority gets on with the business of regulation. The issue of trust in regulators is critical here, but citizens in some jurisdictions are increasingly doubtful about the capacity of regulatory decision-makers to make sound assessments. A 2006 Canadian public opinion survey on biotechnology found that 'four in ten believe the regulatory system is probably either somewhat or very lax, and another 20% say they are uncertain about these systems, which reveals a fairly significant level of concern' (Decima 2006: 5). The fallout from approval of an animal biotechnology product that has adverse consequences may seriously undermine public confidence in regulators. In the health care sector, recalls of heavily promoted and prescribed drugs like Vioxx have had this effect, leading to calls for special labels on drugs that have been on the market for under two years so that consumers know long-term safety is uncertain (Institute of Medicine 2006). A participant in the Regulators focus group commented that 'loss of public faith [in] the ability of the regulatory system to maintain the right checks, controls on an industry . . . could even spill over into loss of public faith in the whole food regulatory system' (Regulator V6). Long-term risks in animal biotechnology raise long-term risks for trust in regulatory systems.

Conclusions

Since the 1970s, biomedical ethics has focused on four key principles: respect for autonomy, beneficence or aiming to do good, nonmaleficence or avoiding harm, and justice (Beauchamp and Childress

2001). While these principles were articulated to guide ethical decisions regarding humans and health care, the values that underlie these principles can be applied to some extent in the novel context of using animal biotechnology to achieve human health goals. In particular, values underlying beneficence, non-maleficence, and justice are applicable to ethical debates about the use of animal biotechnology. In our research, focus group participants were concerned with whether animal biotechnology really would have beneficial outcomes, for both humans and animals. In general, genetic modification of animals was considered more acceptable if many people could benefit, such as using biopharming to produce drugs to treat a widespread disease. Similarly, beneficence towards animals themselves was very important for many participants. Strong concerns for animal welfare were expressed, reflecting values in the humane treatment of animals and avoiding or minimizing harms.

Balancing potential benefits for humans with more immediate harms to animals remains a difficult trade-off. In xenotransplantation, for instance, research animals have tissues and organs removed, other animals, including non-human primates are recipients of those tissues and organs, and they soon die or are euthanized – all in the pursuit of donor organs for humans who, themselves, face death without a suitable transplant. In these cases, focus group participants tended to look for alternatives to animal biotechnology and ask what other medical or policy options are available to address the human health need.

The concern with alternatives to animal biotechnology heightened as focus groups moved away from a clinical medicine context to consider animal biotechnology used to provide healthier foods and environments. Less polluting or omega-3 pigs and BSE-resistant cattle were viewed by many as novel uses of biotechnology that encourage harmful and unsustainable human habits. This reflects a concern with justice and a desire to avoid harms to animals, the environment, and future human generations. Some focus group participants, however, expressed resigned tolerance of such uses of animal biotechnology; in their view, humans have already manipulated the environment to such an extent that genetic alteration of animals is necessary to mitigate further harm.

Finally, values related to justice were reflected in worries that animal biotechnology could exacerbate health inequities, including the abysmal attention to health problems that affects the majority of the world's population, who live in conditions of poverty and ill health. The economic cost of the animal biotechnology research enterprise is high, and applications that are eventually approved for human health purposes may be limited to wealthy individuals or countries that can afford them. Some applications, such as biopharming, are more explicitly aimed at producing drugs at higher volume and lower cost, which could improve access to health care if companies allocate their resources accordingly. Biotechnology applications to reduce zoonotic disease transmission, particularly for endemic diseases like malaria, could also hold promise if ecological concerns are resolved. These approaches may, however, raise the sort of political disputes that have occurred in regard to GM food, where less developed countries object to having GM 'solutions' foisted on them.

It has been said that the future of animal biotechnology is 'only limited by our understanding of the biological system[s] and our imagination' (Prather et al. 2003: 119). The technological ability to modify animals to pursue health goals compels us to think about the complex connections among people, animals, and the environment. As a participant in our Scientists focus group observed, 'Biotechnology [is] a test of how [humans] view themselves in relationship to the world' (Scientist V5). The theme of re-thinking our relationship with animals and the environment emerged in many of the focus group discussions and is a theme that runs through this volume. Engaging in deeper reflection on these issues may be an important collateral benefit of our ever-expanding capacities to use animal biotechnology for many purposes, and can help us elucidate the values at stake and make sound decisions about proceeding down these novel paths.

NOTES

1 See Pharming website: http://www.pharming.com/index.php?act=prod&pg=11.
2 At http://www.oxitec.com.
3 See DengueNet, the WHO dengue surveillance system: http://www.who.int/csr/disease/dengue/denguenet/en/index.html.

References

Auditor General of Canada. 2005. Report of the Commissioner of the Environment and Sustainable Development, Chapter 8, Environment petitions:

Impacts of hog farming. Ottawa: Office of the Auditor General of Canada. http://www.oag-bvg.gc.ca/internet/docs/c20050908ce.pdf.

Bach, F.H. 1998. Xenotransplantation: Problems and prospects. *Annual Review of Medicine* 49: 301–10.

Beauchamp, T.L., and Childress, J.F. 2001. *Principles of biomedical ethics*. 5th ed. Oxford: Oxford University Press.

Burks, A.W. 2008. Peanut allergy. *Lancet* 371: 1538–46.

Canadian Poultry Industry Forum. 2004. Report, submitted to Deputy Minister, British Columbia Ministry of Agriculture, Food and Fisheries, President, British Columbia Poultry Committee, and President, Canadian Food Inspection Agency. http://www.al.gov.bc.ca/avian/CPIF-Avian.pdf.

Canadian Public Health Association. 2001. Animal-to-human transplantation: Should Canada proceed? A public consultation on transplantation. Ottawa: Canadian Public Health Association. http://www.xeno.cpha.ca/english/finalrep/reporte.pdf.

Cyranoski, D. 2008. Sterile mosquitoes near take-off. *Nature* 22(453): 435.

Decima Research. 2006. *Emerging technologies tracking research – Final report*. Prepared for Industry Canada. http://www.ic.gc.ca/epic/site/ic1.nsf/vwapj/Decima-June2006_EN.pdf/$file/Decima-June2006_EN.pdf.

Duguay, F., Katsanis L., and Thankor, V. 2003. The identification of factors linked to the potential acceptance of transgenic pharmaceuticals: An exploratory study. *Health Marketing Quarterly* 21(1/2): 65–89.

Fiester, A. 2006. Why the omega-3 piggy should not go to market. *Nature Biotechnology* 24(12): 1472–3.

– 2007. Casuistry and the moral continuum: Evaluating animal biotechnology. *Politics and the Life Sciences* 25: 16–22.

Golovan, S.P., et al. 2001. Pigs expressing salivary phytase produce low Phosphorus manure. *Nature Biotechnology* 19: 741–74.

Hagelin, J. 2004. Public opinion surveys about xenotransplantation. *Xenotransplantation* 11(6): 551–8.

Institute of Medicine. 2006. *The future of drug safety: Protecting and promoting the health of the public*. Washington, DC: National Academies Press.

Jones, I. 1996. 2010: A pig odyssey. *Nature Biotechnology* 14: 698–9.

Kang, J.X., and Leaf, A. 2007. Why the omega-3 piggy should go to market. *Nature Biotechnology* 25(5): 505–6, followed by a response by A. Fiester.

Karatzas, C.N. 2003. Designer milk from transgenic clones. *Nature Biotechnology* 21: 138–9.

Laible, G. 2009. Enhancing livestock through genetic engineering: Recent advances and future prospects. *Comparative Immunology, Microbiology and Infectious Diseases* 32: 123–37.

Lancet. 2007. Editorial: Legal and illegal organ donation. *Lancet* 369 (9577): 1901.

Mohiuddin, M.M. 2007. Clinical xenotransplantation of organs: Why aren't we there yet? *PLoS Medicine* 4(3): 429–34.

Nature Biotechnology. 2002. Editorial: Will these pigs ever fly? *Nature Biotechnology* 20: 203.

Niemann, H., and Kues, W.A. 2003. Application of transgenesis in livestock for agriculture and biomedicine. *Animal Reproduction Science* 79 (3/4): 291–317.

Pew Initiative on Food and Biotechnology. 2004. Bugs in the system? Issues in the science and regulation of genetically modified insects. Washington, DC. http://www.pewtrusts.org/uploadedFiles/wwwpewtrustsorg/Reports/Food_and_Biotechnology/pifb_bugs_012204.pdf.

Powell, K. 2007. Functional foods from biotech: An unappetizing prospect? *Nature Biotechnology* 25(5): 525–31.

Prather, R.S., et al. 2003. Transgenic swine for biomedicine and agriculture. *Theriogenology* 59: 115–23.

Resnik, D.B. 2004. The distribution of biomedical resources and international justice. *Developing World Bioethics* 4(1): 42–57.

Rudolph, N.S. 1999. Biopharmaceutical production in transgenic livestock. *Trends in Biotechnology* 17: 367–74.

Schubert, D. 2007. Letter to the editor. *Nature Biotechnology* 25(3): 282–3.

Sorenson, J.H. 1990. Ethics and technology in medicine: An introduction. *Theoretical Medicine and Bioethics* 11(2): 81–5.

Sykes, M., d'Apice, A., and Sandrin, M. 2003. Position paper of the Ethics Committee of the International Xenotransplantation Association. *Xenotransplantation* 10: 194–203.

United States Department of Health and Human Services. 2007. OPTN/SRTR annual report, Trends in organ donation and transplantation in the United States, 1997–2006. http://www.ustransplant.org/annual_reports/current/chapter_i_AR_cd.htm?cp=2.

Walkerton Inquiry. 2002. Report. Part 1: Events of May 2000 and related issues and Part 2: A strategy for safe drinking water. Commissioner, The Honourable Dennis R. O'Connor. Ontario Ministry of the Attorney General. http://www.attorneygeneral.jus.gov.on.ca/english/about/pubs/walkerton/.

World Health Organization, 1948. Preamble to the Constitution of the World Health Organization as adopted by the International Health Conference, New York, 19–22 June 1946; signed on 22 July 1946 by the representatives of 61 states (Official Records of the World Health Organization, no. 2,.

p. 100) and entered into force on 7 April 1948. http://www.who.int/about/definition/en/print.html.
– 2004a. Avian influenza A(H5N1) – update 31: Situation (poultry) in Asia: Need for a long-term response, comparison with previous outbreaks (2 March 2004). http://www.who.int/csr/don/2004_03_02/en/index.html.
– 2004b. Global strategy on diet, physical activity and health. http://www.who.int/dietphysicalactivity/strategy/eb11344/strategy_english_web.pdf.
– 2009. Cumulative number of confirmed human cases of avian influenza A/(H5N1) reported to WHO (8 April 2009). http://www.who.int/csr/disease/avian_influenza/country/cases_table_2009_04_08/en/index.html.

8 Animal Justice: Understanding the Implications of Animal Biotechnology for Animal Welfare and Animal Rights

LYNE LÉTOURNEAU

In April 2008, People for the Ethical Treatment of Animals (PETA), a prominent animal rights organization, offered a $1 million reward to the first scientist who could produce and bring to market by June 2012 'in vitro' chicken meat. In vitro meat production involves the use of animal stem cells to grow tissue cultures that mimic animal flesh and can be cooked and consumed like meat (PETA 2008 and Sandhana 2006). According to the PETA organizers, in vitro meat production represents a worthwhile application of biotechnology that could potentially spare from horrific suffering billions of intensively raised livestock and fish. As the organization explained, since 'many people continue to refuse to kick their meat addictions, PETA is willing to help them gain access to flesh that doesn't cause suffering and death' (PETA 2008).

Surprisingly, the decision to sponsor this prize sparked intense controversy within PETA's ranks (Schwartz 2008). One might think the disapproval was founded on the concern that such technologies reinforce an instrumental view of animals as objects for human use and consumption, an attitude we should be striving to change (see Singer 1992: 732). However, the reason underlying the criticism, it was later determined, was unrelated to any concern for animals. In fact, the opposition was connected with the abhorrence many people feel towards the eating of meat. Indeed, internal critics of the PETA campaign were expressing a complex food taboo with deeply seated historical roots, one which views the eating of animal flesh as repugnant (Schwartz 2008).

The case of PETA's in vitro meat campaign is interesting. It illustrates that, contrary to widespread belief, a strong commitment to animal rights does not necessarily entail the rejection of all animal biotechnology applications, just as arguments against the genetic engineering of animals are not based always or entirely on concerns about the welfare of animals. Elsewhere, I have suggested that the notion that animals are members of the moral community applies imperfectly to the debate on animal biotechnology (Létourneau 2005). Following on this idea, my aim in this chapter will be to further this argument by showing the limits of contemporary animal rights theory with respect to debates over the moral acceptability of practices such as animal cloning, genetic engineering, and xenotransplantation (the transfer of genes between humans and animals). My objective is to show how animal welfare ethics alone is insufficient to develop a critique of animal biotechnology, and that other lines of arguments will likely be required in order to substantiate such a condemnation.

My analysis has three parts. The first involves a quick overview of some of the dominant currents within contemporary animal welfare ethics. I will then move to a discussion of data gathered from the Animal Justice (AJ) focus group surveyed as part of this study.[1] In particular, I will examine how principles developed by the animal rights movement are expressed by the AJ group in response to questions concerning specific animal biotechnology applications. Two findings from the group stand out in this regard: first, on the one hand, animal justice advocates express strong disagreement with most applications of animal biotechnology, and second, at the same time, they show detachment towards the idea of gene transfer generally. I will argue that these two positions, though seemingly contradictory, are actually consistent with the world view proposed by the framework of contemporary animal welfare ethics. In the concluding section, I will reflect on the nature and extent of the contribution that contemporary animal welfare ethics can make to the issue of the moral acceptability of animal biotechnology. Animal rights theory can indeed be helpful in developing measures to protect animals used in biotechnology research, but, because it is ambivalent on the broader questions of genetic integrity, gene mixing, and species boundaries, and so on, it does not provide the conceptual apparatus needed to address this set of issues.

Contemporary Animal Welfare Ethics

Numerous works over many centuries have nurtured the discussion on animal ethics. The following account will focus on the last thirty years or so of developments within this area of philosophical inquiry. Since the 1970s, proponents of animal welfare ethics have called into question the keeping and killing of animals for human purposes. The leading works in the field have proposed a range of arguments and reforms in response to the following three basic moral questions: First, is the use (i.e., the keeping and killing) of animals for human purposes morally acceptable? If it is, then the second question asks, what, if any, limitations should be placed on the use of animals? The third question, related to standards of treatment, asks what sort of treatment should be accorded to animals in the actual context of their use (Francione 1995: 172)? The first two questions refer to a first level of moral reflection, that of animal use per se. The third one pertains to a second level of moral reflection – that of how animals are treated.

As an area of philosophical inquiry, contemporary animal welfare ethics has focused primarily on the first level of moral reflection. Delving into the issue of the moral status of animals, philosophers and other scholars have looked at whether animals are worthy of moral consideration in their own right. They have asked whether animals matter at all from a moral point of view, and if they do, what do we owe them?[2] By and large, these discussions have centred on the views put forward by philosophers Peter Singer and Tom Regan. The views articulated by Singer and Regan are both illustrative of a position that grants moral status to animals. Both philosophers recognize that animals are worthy of moral consideration, and both extend to animals 'a principle of equal consideration,' implying that the needs and interests of animals should be considered equally to those of humans.

From this point on, however, Singer and Regan diverge as to how the principle should be applied. In his book *Animal Liberation* (Singer 1990), Singer argues for an approach based on utilitarianism, a doctrine he endorses throughout his professional philosophical work (Varner 2002: 154). As a moral theory of right and wrong action, utilitarianism holds that we ought to act so as to bring about the greatest balance of good over evil for everyone affected. Singer argues that it is necessary, when assessing the consequences of our actions, to take into account the interests of every being affected and to give these interests the same weight as the similar interests of any other being. His notion of moral action requires equal consideration of the interests of both humans and animals.

In *The Case for Animal Rights* (Regan 1983), Regan defends a position founded on a very different structure – that of basic moral rights that are shared equally by all individuals who possess an inherent value, be they human or non-human. These basic moral rights cannot be violated or sacrificed for the benefit of others. They erect protective fences around animals and impose limits on what humans can do to them. Central to Regan's view is the idea that animals are never to be treated as a means to human ends, however good those ends may be (Regan 1983).

Because Singer's view is utilitarian and Regan's is rights-based, 'a utility-versus-rights debate has clamoured loudly in the literature' (DeGrazia 1999: 112) as a mirror image of the vigorous opposition in moral theory generally between consequentialism and deontology. This central debate within animal ethics has set in opposition two competing conceptions of what is owed to animals as bearers of moral status: equal consideration of animal *interests* as an integral part of the utilitarian calculation, or stringent safeguarding of animal interests through the recognition of basic moral *rights*. In practical terms, the debate is important for animals, since, in contrast with Regan's uncompromising abolitionist stance (Regan 1983: 330–400), Singer cannot avoid entertaining the possibility that harm caused to animals by a certain act might – in some limited circumstances – be outweighed by a greater positive outcome (see, e.g., Singer 1992: 731–2; and Sztybel 2006: 174–6).

But stepping back for a moment to the earlier question of moral status, how is it that Singer, Regan, and others arrive at the view that animals are worthy of moral consideration?

Within Singer's utilitarian approach, sentience is determinative of moral status. To be sure, utilitarianism aims to maximize welfare, understood in its etymological sense, as philosopher L. Wayne Sumner points out, as 'the condition of faring or doing well' (Sumner 1996: 1, 3). In this sense, writes Sumner, 'welfare attaches pre-eminently to the lives of individuals, and a [being's] welfare is more or less the same as her well-being or interest or (in one of its many meanings) her good' (1). For the same reason, DeGrazia defines an interest as 'something that figures favourably in the welfare, good, or prudential value profile . . . of a particular individual. Moreover, X is in A's interest – and A *has an interest in* X – if and only if X is one of A's interests in the sense just defined' (DeGrazia 1993: 17–18). Now, the basic attribute required for having one's welfare – and thus one's individual interests – taken into consideration is sentience (DeGrazia 1996: 226–31; Singer 1990: 7–8). Since animals are sentient, Singer argues, there is no moral justification for refusing to take their welfare into account.

Regan is also committed to the protection of welfare as a basic good, and advances the 'subject-of-a-life' criterion as its basis: all those who are subjects-of-a-life are worthy of equal moral consideration because they have inherent value. In Regan's view, the basic similarity between human moral agents and what he calls 'moral patients' – those who lack the ability to formulate and bring to bear moral principles when deliberating on a course of action[3] – is that each of them is the experiencing subject-of-a-life that is better or worse for them regardless of their utility to others or their being the object of another's interests. Each one wants and prefers things, believes and feels things, recalls and expects things. All of these dimensions of life, including pleasure and pain, enjoyment and suffering, satisfaction and frustration, continued existence or untimely death, make a difference to the quality of one's life as experienced by her or him as an individual. In short, each moral agent and moral patient has an 'experiential welfare' – an experiential life that fares well or ill, depending on what happens to or is done to or for them. Since this is true of normal mammalian animals of at least one year of age (Regan 1983: 1–120), they too must be regarded as subjects-of-a-life with an inherent value of their own (Regan 1983: 241–5, 262; Regan 1991: 85–6).

Besides sentience and the subject-of-a-life criterion, other justifications proposed for assigning moral status to animals include philosopher Evelyn B. Pluhar's notion of the moral significance of sentient 'conative' beings, that is, those who have purposes they want to fulfil (Pluhar 1995). According to Pluhar, although not all animals are consciously purposive beings, 'we do appear to be empirically warranted in attributing consciously purposive behavior to mammals' (ibid.: 257). Other philosophers, such as Mary Ann Warren (Warren 1997), offer a multi-criteria approach in keeping with which different levels of status are hypothesized, 'suggesting that some entities have partial moral status while others have full moral status or no moral status' (Hale 2006: 354). What is interesting to note among these viewpoints is their common way of arguing that animals deserve moral status. Indeed, all follow this simple pattern: humans must afford moral consideration to animals because animals possess characteristics or attributes that make them morally considerable in their own right (Hale 2006: 349). As DeGrazia underscores, 'the justification for assigning moral status (or some level of moral status) to a being, according to nearly all of the animal ethics literature, appeals to [certain of] the individual's *properties*' (DeGrazia 1999: 126).

From this overview of contemporary animal welfare and rights ethics, six key features emerge. For most proponents, the principles of animal welfare ethics

1 offer a critique of animal use, understood as the keeping and killing of animals for human purposes;
2 address the issue of the moral status of animals;
3 argue that animals have interests and rights based on their possessing certain attributes, such as sentience, being the subject-of-a-life, or conation;
4 suggest that the interests of humans do not weigh more, as a rule, than the interests of animals, but rather that comparable or similar interests should be given equal moral weight regardless of who is the bearer of those interests;
5 focus on welfare as a basic good or value to be protected;
6 propose varying levels of protection for animals, for example, a utilitarian balancing of interests or complete and fundamental protection of rights.

Keeping in mind these key features of contemporary animal ethics, the following section will move to a discussion of the views articulated by participants in the Animal Justice (AJ) focus group. The purpose is to see how the principles of animal ethics are applied to the cases involving the use and treatment of animals in biotechnology settings.

Views from the Animal Justice Focus Group

As one might expect, the participants in this group express a fairly unanimous condemnation of the use of animals (or, at least, some animals) in biotechnology research and food production applications. It is an opposition, moreover, based on a classical rights-based view of the moral status of animals – one derived from their condition of sentiency. What is also significant, however, is how the attitude towards animals articulated by the group leads to ambivalence towards the practice of gene transfer and other aspects of genetic engineering that may impinge on the genetic integrity of species. Throughout the discussion, my aim is to show that such a twofold position, though apparently inconsistent, coincides completely with an ethical framework that views humans and animals as having equal moral status.

If we would not do this to humans, why are we doing it to animals? In framing their concerns about animal biotechnology, the AJ focus group participants make use of one the fundamental premises underlying animal-based research, namely, the idea that animals are biologically as well as morally similar to humans. As one participant expressed it, 'if animals were so different from [humans] that there was nothing to be learned that would benefit [us] by . . . using them in experiments, we wouldn't do it. There would be no point' (Animal Justice stakeholder V2).

It is indeed well established that animals, including all vertebrates and some invertebrates such as cephalopods, are capable of experiencing physical and mental pain, just as humans do (see, e.g., DeGrazia 1996: 105–23; Smith and Boyd 1994: 45–77). As sentient beings, their welfare is affected by the actions of others. Humans and animals thus share an interest in the avoidance of pain and suffering (DeGrazia 1993: 18). According to the group, all sentient beings have 'an equal right to have [their] interests considered' (Animal Justice stakeholder V2).

Although reminiscent of Singer's utilitarian viewpoint, this view does not sit well with the consequentialist outlook. In consequentialist theory, what is morally right or wrong under the utilitarian doctrine is determined strictly in terms of consequences. However, the AJ focus group participants are against 'any animal being used if it's to its physical or mental detriment' (V1), regardless of consequences. As this participant makes clear in relation to the use of animals for biomedical research: 'I've never said that I don't think that there aren't answers that come out of animal-based research. I think there are . . . With all the millions, if not billions, of animals that have been used to find answers, if nothing had been found from it, you know, that would be pretty incredible. But I subscribe to the idea that we don't have a right to be doing what we are doing.'

Indeed, the members of this group generally express a position much closer to Regan's, that is, one that views violating the interests of animals as intrinsically wrong, no matter how many benefits might be gained for humans or animals. This aligns more closely with a deontological approach to ethics, in which actions are morally right or wrong to the extent that they coincide with or contradict our moral duties. For this reason, their attitude towards animals qualifies as a rights-based position.

Extending from this rights-based view is another idea in animal ethics, also expressed by the AJ focus group participants, that though humans may wish to use animals for things such as biomedical research, we cannot consistently do so if we cannot at the same time accept the implication of analogous use of humans for biomedical research (Rachels 1999: 130). If animals are morally as well as biologically similar to humans, then it is inconsistent to use animals willingly for research purposes while refusing to consider using humans, particularly those such as severely brain-damaged infants or the cortically dead, who are less intellectually developed than many non-human animals, for the same purposes (Singer 1992; Dombrowski 1997; Frey 2001; Singer 1990: 81–5; Sztybel 2006). As Singer argues in his version of the 'argument from marginal cases,' human subjects who have only the mental capacities of animals should also be used for experiments, since they would be more suitable biologically for medical research concerning humans and the results more valid (Sztybel 2006: 175). If one is not prepared to go down that road, then one should refrain from using animals. This is essentially the view that AJ focus group participants put forward. The basic principle at work here, extending from this rights-based view of morality, is that 'similar cases should be judged similarly' (DeMarco 1996: 55–6; see also Russow 2002: 65).

A Twofold View of Animal Biotechnology

In view of the overarching attitude towards animals and their moral status expressed by the AJ focus group, how do they view the specific case of animal biotechnology? For the most part, AJ focus group participants cited the 'exploitation of animals' as their main concern with respect to animal biotechnology. In particular, they reject all applications of animal biotechnology on the basis that they interfere with the interests of animals. However, because it cannot be reconciled with a proper understanding of welfare, the notion of genetic integrity does not, in the view of this group, constitute an animal interest. This explains why certain applications of genetic engineering (e.g., making animals more fit for their environment) gave rise to no particular negative reactions from AJ focus group participants. In addition, group members formulated no disapproval as such against the transfer of genes across the kingdoms of plant and animal or human and animal life. This leaves the door open to practices, for example, brain tissue transfers between humans and primates, that members of other focus groups found to be quite questionable. For the AJ group, though, moral status is not determined by species membership, but is established on the basis of sentiency, along with the interests that are constitutive of each individual's welfare. Moreover, the responses from this group emphasize that what is important are the duties that we owe to sentient creatures once they exist, regardless of their origin. In sharp contrast, participants from other focus groups, who express a very different attitude towards animals (i.e., anthropocentric), strongly condemned applications of animal biotechnology they perceived as threatening the divide between humans and animals. As I will argue later on, such detachment towards gene transfer generally is related to the attitude towards animals adopted by AJ focus group participants and follows from their commitment to contemporary animal ethics.

Animal Exploitation and Animal Interests

While the potential benefits to human health, economic development, environmental protection, and so on that can result from animal biotechnology are generally recognized by members of the AJ focus group, they nonetheless feel that such practices could never be supported because of the high level of animal exploitation involved. It was noted as one concern, for example, that cloned and transgenic animals lead confined lives in which they are not allowed to 'go about life on [their] own' (V3). As one person commented about cloned calves: 'There's just nothing normal about it. Everything is just totally Frankenstinian, in terms of the animals can't – they can't live a normal life. They are kept away from other animals; they're kept behind . . . barriers. They'll never go outside. They'll . . . never have a natural life, even assuming they do survive' (Animal Justice stakeholder V1).

The waste of animal lives through biotechnology research is another concern. To create a new transgenic line of mice, for instance, requires euthanizing many non-transgenic mice used at different stages along the way for breeding, testing, and embryo donation (Dennis 2002: 102–3). Cloned and transgenic animals also experience a high rate of mortality and birth defects. Among the small percentage of animal clones alive at delivery (about 5 to 25 per cent of cloned embryos), 14 per cent suffer from problems that impact negatively on their health and welfare, including fetal overgrowth and malformations. Many of these animals do not survive past a few weeks of age (Bettayeb 2008: 52).

The members of the AJ focus group note the infringement in animal biotechnology of three different types of animal interests: the interest in not being killed or caused to die prematurely, the interest in living a natural life, and the interest in retaining bodily integrity. In chapter 3 of this volume, Mickey Gjerris refers to the concept of animal welfare as having both a narrow and a broad definition. Within the narrow definition, Gjerris explains that only the health and subjective experience of the animal have ethical importance. From the broader perspective, however, animal welfare is also about the extent to which the animal is allowed to fulfil its species-specific potential, regardless of its subjective experience. Within this schema, the AJ focus group participants could be said to adhere to a narrow definition of animal welfare. This is because they concentrate primarily on the *subjective* experience of animals.

In the case of animals' interest in not being killed or caused to die prematurely, for example, DeGrazia writes that 'all beings with interests have an interest in life, [which is] a precondition for the satisfaction of almost every other possible interest' (DeGrazia 1993: 23; see also Pluhar 1995: 281). As a prerequisite for successful experiential well-being, remaining alive thus falls within the ambit of a narrow, subjective outlook on animal welfare.

Similarly, the interest in living a *natural* life is construed in terms of the subjective experience of animals. Philosopher Bernard Rollin designates under the name of *telos* the idea that animals 'have natures, genetically based, physically and psychologically expressed, which determine how they live in their environments' (Rollin 1995: 159). Above it was noted how AJ focus group participants are worried about the ability of transgenic and cloned animals to engage in natural, or species-specific, behaviours. Although such language might suggest the espousal of Gjerris's broad definition, their concern relates to the suffering caused to animals when they are denied the ability to follow their instincts or to develop and use their natural adaptations and capabilities. In a further illustration of this, when asked to comment on the example of battery hens that have been genetically altered to have their urge to nest removed, AJ focus group participants expressed no disquiet over such an application that could help to make animals more fit for their intensive rearing environment, an answer that indicates again a preoccupation with subjective experience.[4]

The Problem of Genetic and Bodily Integrity

Some might perceive this response to the example of the hens as a refusal by the group to engage with the issue of transgenesis itself.

I would suggest, rather, that such a position extends naturally from the conceptual apparatus provided by contemporary animal welfare ethics as it has developed so far. To take the notion of *telos*, for example, note how the following passage from Rollin in his book *The Frankenstein Syndrome* defines the concept from the standpoint of animal ethics:

> [My] notion [of telos] has been misunderstood in very serious ways. It has been interpreted by critics of genetic engineering (and sometimes its supporters) as meaning that my position asserts that an animal's telos cannot be altered without violation, and thus, according to my account, all genetic engineering is wrong. I did not, of course, make that claim. *What I did assert was that given an animal's telos, and the interests that are constitutive thereof, one should not violate those interests. I never argued that the telos itself could not be changed.* If the animals could be made happier by changing their natures, I see no moral problem in doing so (unless, of course, the changes harm or endanger other animals, humans, or the environment). *Telos is not sacred; what is sacred are the interests that follow from it.* (Rollin 1995: 171–2, emphasis added)

So what, then, of the notion of genetic integrity, the idea that one has an interest in having one's genome left intact? We have seen how protecting welfare is central to animal welfare ethics, and how the welfare of individual animals, also understood as the quality of their lives, correlates with the protection of interests. In particular, any harm to the relatively similar interests of humans and animals requires equal consideration. With respect to animal biotechnology, the question is thus: Do animals have an interest in genetic integrity or, otherwise stated, in their genome being left intact?

I would argue, drawing on the work of philosopher Peter Sandøe, L.W. Sumner, and others, that within the framework of animal welfare ethics as currently defined, such an idea does not stand up to scrutiny. As Sandøe writes, whether or not we respect genetic integrity 'will have consequences concerning which animals come into existence and which don't'; but, he adds, respect for genetic integrity will not affect '*whether* animals benefit or are harmed' (Sandøe et al. 1996: 118, my emphasis). Whether one's genetic make-up is intact or altered is a condition an animal or person is born with. It is strictly a fact about one's characteristics as a living being – just like a physical or mental disability. Unlike access to food, water, shelter, companionship, or freedom, genetic integrity is not about states of the world that make an animal's life go better or worse; it is intrinsically tied to an animal's self. For that reason, it involves nothing close to an interest that would figure as a constituent of an animal's welfare.

Some, such as animal welfare scientist Michael Appleby, have suggested otherwise, arguing that to genetically modify an animal's instinct for behaving in ways typical to its species 'is *de facto* a compromise of the welfare of the animal' (summarized in Thompson 2008: 69). This view, which finds its roots in environmental ethics, defends an objective and teleological theory of welfare based on the idea of proper functioning, in which welfare depends on the development of all of an animal's essential characteristics as predetermined by some external source (see Sumner 1996: 72). This understanding relates to the broader definition of animal welfare described by Gjerris.

I would argue, though, that this does not constitute a plausible theory of the nature of welfare. In his book *Welfare, Happiness and Ethics* (Sumner 1996), Sumner notes that 'welfare assessments concern what may be called the prudential value of a life, namely, how well it is going *for the individual whose life it is*' (20). In this view, subject-relativity 'is a key ingredient in our concept of welfare' (20). For Sumner, objective theories of welfare such as Appleby's, which do not incorporate an internal reference to the bearer or take the point of view of the subject, are inadequate (27, 42–3). In particular, conceptions of welfare built on the idea of proper functioning 'conflate prudential and perfectionist value: they are really theories about the latter rather than the former' (78–80). In sum, only the narrower perspective on animal welfare defined by Gjerris and adopted in contemporary animal ethics reflects a proper understanding of the nature of welfare.

This brings us to the third remaining interest mentioned by the AJ focus group participants: the interest in retaining *bodily* integrity. If respect for genetic integrity does not suit the narrative of contemporary animal ethics, does the same conclusion follow for the preservation of bodily integrity? Based on the previous discussion, the answer would appear to be yes, if what is meant by integrity implies a state of being 'normal' and without deformities or debilitating physiological problems, presented as an objective criterion. The problem is that it is unclear whether AJ focus group participants view bodily integrity strictly in relation to its impact on prudential welfare or also as an objective criterion. To some extent, they allude to both senses in their thinking, referring to abnormality as both a cost to the animal and also as not being the way animals are supposed to be.

To summarize, according to AJ focus group participants, animal biotechnology is problematic not because there is anything intrinsically wrong with the cloning or genetic engineering of animals, but because it violates the interests of experimental animals used in the creation of animal clones and transgenic animals and those of transgenic and cloned animals once they exist. For the reasons given, they express no strong adverse reaction to the prospect of using genetic engineering to make animals more fit for their environment. Centring essentially on the subjective experience of animals, AJ focus group participants articulate a way of thinking about animal biotechnology that is continuous and consistent with their animal rights position.

Gene Transfer across the Kingdoms

The ambivalence shown by animal welfare advocates towards animal biotechnology becomes even more apparent in the context of discussions that took place in this study over the acceptability of transferring genes across the plant, animal, and human kingdoms. Whereas the other focus groups in the study were typically favourable to animal cloning and genetic engineering,[5] this was one example where their voices joined unanimously with those of the AJ group in rejecting this type of animal biotechnology application as completely unacceptable. The case at hand concerns the example of genetically modifying a primate to incubate human brain tissues. Whereas AJ focus group participants condemned the application for the fact that it opens up 'a new area of exploitation' (V3), participants from other focus groups attacked the application as fundamentally offensive.

What is it about this application that triggered such unanimous opposition from the other groups? Their objections seem to have been based on strong feelings of repulsion or abhorrence (the 'yuk factor,' as termed by Midgley [2000]) that can be traced to a series of concerns related to identity. Indeed, in contrast with AJ focus group participants, several members of other focus groups expressed disgust at the mention of genetically modifying a primate to incubate human brain tissues. As one of them asserted, 'It gave me the creeps' (Regulator V7).

One potential explanation for such abhorrence is found in the biological, or evolutionary, proximity that exists between primates and human beings. The fact that primates are 'so close to us' engenders the anxiety that mixing human with primate genes may threaten our identity as beings separate from animals. This argument has been evident in debates over xenotransplantation (the transplantation of living cells, tissues, or organs from animals to humans), in which it is argued that xenotransplantation is morally unacceptable because it promotes a conception of human identity that does not take into account the significance of the body in the formation of one's sense of self (see, e.g., Advisory Group on the Ethics of Xenotransplantation 1996: 68; Comité consultatif national d'éthique 1999; Nuffield Council on Bioethics 1996: 104; and Schroten 2001: 9). According to these critics, xenotransplantation conveys a reductionist attitude that views the body as instrumental, that is, as excluded from the essence of what it means to be human. In addition, it proposes a dualist vision that separates and opposes the body to all other components of human identity, the body being subordinate to these other elements.

The focus group participants took this argument further, expressing a view that since the brain in particular represents the essence of self, to attempt to use primates to incubate human brain tissue would involve mixing different orders of creation, which could potentially produce 'confusion,' that is, a blurring of identity where, on the one hand, a primate might acquire human thought processes and as result become partly human-like and, on the other hand, a human might penetrate the foreclosed land of the primate mind and thus partly develop into an animal. The perceived threat to identity resides in the possible manifestation of the double-sided 'confusion' described.

This uneasiness reverberates with ancient proscriptions against bestiality, in which the act is portrayed early on in the history of European morals as one of the most serious and degrading sins and the worst of sexual crimes (Gazzaniga 1987: 49–50; Thomas 1983: 39). In the Middle Ages, the punishment for bestiality could involve both parties being burned alive (Evans 1906: 147). Here again, the concern is with the preservation of human identity. What we fear, writes Stout, 'when the shepherd's lust takes his flock as its object,' is that the shepherd 'has become too beastly to maintain a firm grip on his social identity' (Stout 2001: 152). Bestiality, in brief, imperils 'our unambiguous status as human beings' (ibid.: 151).

Indeed, going back to the Greeks, humans have long sought for ways to affirm the unique morality of humans in relation to animals. This line of philosophy has lent support to an anthropocentric outlook, well established in the West, that views the interests of humans as morally more important than the interests of animals, or of nature in its totality (De Roose and Van Parijs 1991: 23). To this day, many criteria have been offered to justify a human-centred ethic, including the possession of a soul, language use, the capacity to reason, the capacity for moral agency, and so on (see, e.g., Pluhar 1995: 10–57). Although these criteria have been forcefully criticized in contemporary animal ethics, anthropocentrism still remains the dominant viewpoint within Western society.

For the most part, the focus group members in the study, like most people in the West, hold an anthropocentric view of the world. As one person said, 'I do distinguish animals from the human population. I think that's where the division is' (Agricultural Producer stakeholder V4).

What is more, participants are approving of cases where animals are treated differently from humans, such as selective breeding, which is 'a good thing' for animals, but unacceptable for humans, or germ-line modification, which is ruled out from the start for humans but envisioned for animals. Within this division, however, are gradations. As another participant asserts, 'I am less concerned about a mouse than I would be over, perhaps, a primate or some other type of higher life form' (Agricultural Producer stakeholder V1). The notion of gradations is consistent with anthropocentrism because the standard against which one's level of ethical concern is measured still remains kinship with members of the human community.

In view of these considerations, it is likely that the abhorrence expressed by these focus group participants at the incubation of human brain tissue in primates correlates with the same primal aversion to bestiality: by threatening the divide between humans and animals, it calls into question our separate identity as human beings. If identity inheres in the brain – as participants believe – then being unambiguously human entails possessing a brain that has developed inside a human body. What's more, being unequivocally one's self requires having a brain that has grown inside one's own body since birth (see Stout 2001: 151). Such experiments risk blurring the line between humans and nonhumans, endangering both social and personal identity.

Whereas members from other focus groups view humans and animals as distinct classes of being, AJ focus group participants draw no firm line between them, viewing humans as one part of a community of living organisms. On the issue of xenotransplantation, they consistently return to the principles of sentiency and subjective experience as the basis for judging moral acceptability, rather than the protection of species boundaries. As one person commented, 'To me, if you're sentient, there's ethical obligations. So I don't care if you're part pig, part human, part goat, whatever you are, there's ethical obligations by virtue of your sentience' (Animal Justice stakeholder V2).

Outside of the ethical domain circumscribed by sentiency and attendant moral status, AJ focus group participants do admit that there are 'all sorts of other kinds of concerns' (V2). They cite the unknown consequences of tampering with nature and mixing genes, as well as the problem of human arrogance in the practice of science, which is falsely represented as being able to solve all problems. They are also aware of arguments about the dangers of playing God, and the 'slippery slope' that is encountered as biotechnology pushes the boundaries of acceptable genetic intervention to further and further extremes. Such arguments are widespread commodities on the market of ideas that surrounds social controversy on genetic engineering (see, e.g., Comstock 2000; and Richard and Létourneau 2006). Ultimately however, the objections of the AJ focus group participants do not rest on these arguments. As animal advocates, their standpoint sits firmly within the ethical domain drawn by their rights-based attitude towards animals.

Conclusions

In keeping with contemporary animal ethics, what primarily matters for AJ focus group participants are the duties owed to animals as sentient beings worthy of moral consideration. The focus of their attention relates strictly to the issue of moral status. Within this view, how we ought to treat individual animals – or other humans for that matter – does not rely on whether an animal is transgenic or not, and, in the case of genetically engineered animals, is not determined by the 'origin' of an animal's genetic make-up – that is, whether its genome contains a gene taken from another species or not – which is totally immaterial. Following contemporary animal ethics, the only issue that is morally relevant to them is whether genetically modified animals are sentient and, for this reason, full members of the moral community deserving due consideration and respect.

For critics of biotechnology who dislike the very idea of genetic engineering, that is, who maintain that there is something objectionable in the very process of genetically modifying animals, the implications of contemporary animal ethics with regard to animal biotechnology will seem insufficient. The centrality of welfare within animal ethics is not congruent with the emphasis on genetic integrity that underlies criticisms aimed at genetic engineering per se. Similarly, the emphasis on moral status and the animal justice view of a moral community that extends beyond humans offers no fertile ground for challenging gene transfers between species or between the kingdoms of human, plant, and animal life. Animal rights arguments can provide some protection against animal biotechnology, but only insofar as the experiential well-being of animals is interfered with.[6] Were advances in animal biotechnology to succeed in minimizing the keeping and killing of experimental animals in the creation of animal clones and transgenic animals, and the keeping and killing of animal clones and transgenic animals for human purposes, then contemporary animal ethics would lack the required analytical tools to express disapproval of a number of applications.

To oppose animal biotechnology applications like PETA's in vitro meat, one must move beyond the narrative of contemporary animal ethics. DeGrazia has raised the prospect of a critique of animal biotechnology based on the concept of *respect* for genetic integrity. He asks whether 'considerations of respect,' not for autonomy necessarily, but of some other kind, might define a sense of moral duty that transcends concerns about harming, benefiting, and other related concepts (DeGrazia 1999: 128). Such a device, which draws on virtue ethics,[7] might well be relevant in the context of animal biotechnology, but it is an area that remains as yet undeveloped.

Alternatively, one can examine the anthropological roots of one's sense of abomination, such as repulsion to eating meat. But the significance of the abominable to moral deliberation, including what to do with the categories of our cosmology and social structure that give rise to one's sense of revulsion in the first place, awaits a clear understanding (Stout 2001). Consequently, until that time when plausible lines of argumentation are developed to enrich the conceptual apparatus of ethical theory, contemporary animal ethics will continue to apply imperfectly to the debate on animal biotechnology.

NOTES

1 See the introduction to this volume for a discussion of the focus group research used in this study. I wish to express my sincere thanks to all the members of the Animal Justice focus group. The discussions that took place

were extremely informative and were a great help in writing this chapter. I apologize for any misrepresentations that may have unintentionally been introduced into the description of the views provided here.

2 According to philosopher David DeGrazia, a being has moral status *if and only if* that being's interests have some moral weight independent of the way in which protecting those interests redounds to the benefit of others (DeGrazia 1993: 25). Otherwise, although it might be recognized that this being's interests matter altogether, the consideration afforded to them only follows from the pursuit of an ulterior, beneficial end.

3 Infants, the mentally infirm, and animals are paradigm examples of moral patients (Regan 1983: 152–3).

4 That is, the subjective experience of such transgenic animals. But AJ focus group participants would still oppose this application on the basis that it violates the interests of experimental animals used in the creation of those transgenic animals.

5 The opinions of the other groups towards animal biotechnology were generally favourable, but did vary from strong endorsement to weak support depending on the type of application. The strongest levels of support were given for genetically modifying a mouse to increase its susceptibility to cancer and genetically modifying a cow to produce insulin in its milk. Not surprisingly, these applications are supported because of their potential for contributing to the prevention and treatment of human diseases (i.e., cancer and diabetes), as well as the advancement of scientific understanding. Another rationale used by the focus groups relates to the continuity of these practices with historically sanctioned forms of animal use. This is known as the *argument from tradition*. Casuistic by nature, the argument from tradition rests on the principle that an action 'A' is acceptable if and only if it compares favourably with another action 'B,' which has been judged as ethically acceptable by society for many years already. In the case of laboratory mice, since they have been used in these settings for many years and are adapted to a laboratory environment, genetically modifying them for cancer research does not constitute a new form of animal use. Along similar lines, focus group members maintain that genetically modifying a cow to produce insulin in its milk lies 'within the sort of scope of what [cows] have historically been used for.'

6 Since the adoption in England of the first anti-cruelty statute in 1822, the number of laws and regulations aimed at protecting animals has grown considerably at the international level – mostly in Europe. According to political theorist Robert Garner, these measures have done little to challenge the dominant anthropocentric approach to animal welfare, in which protective laws are enacted only to the extent they do not place serious impediments to the interests of humans in using animals for their own

benefit (Garner 1993: 234). Others, however, are more optimistic about the potential of animal rights theory to achieve more robust limits on human use of animals. See, for example, legal scholar Gary L. Francione's strategy for cumulative and incremental reforms that could lead eventually to the abolition of animal use in research and food production, an approach that has much relevance for animal biotechnology (Francione 1996: 190).

7 Virtue ethics connects the manner in which we act with building those traits of character that make one a good person (see, e.g., Rachels 1999: 175–93).

References

Advisory Group on the Ethics of Xenotransplantation. 1996. *Animal tissue into humans*. London: HMSO (Chair, I. Kennedy).

Bettayeb, K. 2008. Allons-nous manger des clones? *Science & Vie* 1088: 48–53.

Comité consultatif national d'éthique. 1999. Avis sur l'éthique et la xénotransplantation. No. 61. http://ccne-ethique.fr/docs/fr/avis061.pdf.

Comstock, G.L. 2000. *Vexing nature? On the ethical case against agricultural biotechnology*. Boston: Kluwer Academic Publishers.

DeGrazia, D. 1993. Equal consideration and unequal moral status. *Southern Journal of Philosophy* 31: 17–31.

– 1996. *Taking animals seriously: Mental life and moral status*. Cambridge: Cambridge University Press.

– 1999. Animal ethics around the turn of the twenty-first century. *Journal of Agricultural and Environmental Ethics* 11: 111–29.

DeMarco, J.P. 1996. *Moral theory: A contemporary overview*. Boston: Jones and Bartlett Publishers.

Dennis, M.B., Jr. 2002. Welfare issues of genetically modified animals. *ILAR Journal* 43(2): 100–9.

De Roose, F., and van Parijs, P. 1991. *La pensée écologiste: Essai d'inventaire à l'usage de ceux qui la pratiquent comme de ceux qui la craignent*. Brussels: De Boeck-Wesmael.

Dombrowski, D.A. 1997. *Babies and beasts: The argument from marginal cases*. Urbana: University of Illinois Press.

Evans, E.P. 1906. *The animal prosecution and capital punishment of animals*. London: William Heinemann.

Francione, G.L. 1995. *Animals, property, and the law*. Philadelphia: Temple University Press.

– 1996. *Rain without thunder: The ideology of the animal rights movement*. Philadelphia: Temple University Press.

Frey, R.G. 2001. Justifying animal experimentation: The starting point. In *Why animal experimentation matters: The use of animals in medical research*, ed. E. Frankel Paul and J. Paul, 197–214. New Brunswick, NJ: Transaction Publishers.

Garner, R. 1993. *Animals, politics and morality*. Manchester: Manchester University Press.

Gazzaniga, J.L. 1987. La sexualité dans le droit canonique médiéval. In *Droit, histoire et sexualité*, ed. J. Poumarède and J.-P. Royer, 41–54. Lille: Publications de l'espace juridique.

Hale, B. 2006. The moral considerability of invasive transgenic animals. *Journal of Agricultural and Environmental Ethics* 19: 337–66.

Létourneau, L. 2000. Animal protection law in Great Britain: In search of the existing moral orthodoxy. Doctoral thesis, University of Aberdeen (unpublished).

– 2005. The regulation of animal biotechnology: At the crossroads of law and ethics. In *Crossing over: Genomics in the public arena*, ed. E. Einsiedel and F. Timmermans, 173–92. Calgary: University of Calgary Press.

Midgley, M. 2000. Biotechnology and monstrosity. *Hastings Center Report* 30(5): 7–15.

Nuffield Council on Bioethics. 1996. *Animal-to-human transplants: The ethics of Xenotransplantation*. London: Nuffield Council on Bioethics.

PETA. 2008. PETA offers $1 million reward to first to make in vitro meat. http://www.peta.org/feat_in_vitro_contest.asp.

Pluhar, E.B. 1995. *Beyond prejudice: The moral significance of human and nonhuman animals*. Durham, NC: Duke University Press.

Rachels, J. 1999. *The elements of moral philosophy*. 3rd ed. Boston: McGraw-Hill College.

Regan, T. 1983. *The case for animal rights*. Berkeley: University of California Press.

– 1991. The case for animal rights. In *Animal experimentation: The moral issues*, ed. R.M. Baird and S.M. Rosenbaum. Buffalo: Prometheus.

Richard, V., and Létourneau, L. 2006. *Portrait du questionnement éthique entourant la transgenèse des plantes et des animaux*, ed. L. Letourneau, 13–55. Montreal: Éditions Thémis.

Rollin, B.E. 1995. *The Frankenstein syndrome: Ethical and social issues in the genetic engineering of animals*. Cambridge: Cambridge University Press.

Russow, L.-M. 2002. Reasoning. In *Life science ethics*, ed. G.L. Comstock, 33–66. Ames: Iowa State University Press.

Sandhana, L. 2006. Test tube meat nears dinner table. *Wired* 06–21–06. http://www.wired.com/science/discoveries/news/2006/06/71201.

Sandøe, P., Holtug, N., and Simonsen, H.B. 1996. The ethical limits to domestication. *Journal of Agricultural and Environmental Ethics* 9(2): 114–22.

Schroten, E. 2001. Xenotransplantation and moral theology. *Veterinary Sciences Tomorrow* 4: 1–12.

Schwartz, J. 2008. PETA's latest tactic: $1 Million for fake meat. *New York Times*, 21 April, http://www.nytimes.com/2008/04/21/us/21meat.html.

Singer, P. 1990. *Animal liberation: A new ethics for our treatment of animals*. Rev. ed. New York: Avon Books.

– 1992. Xenotransplantation and speciesism. *Transplantation Proceedings* 24(2): 728–32.

Smith, J.A., and Boyd, K.M. 1994. *Lives in the balance: The ethics of using animals in biomedical research*. Oxford: Oxford University Press.

Stout, J. 2001. *Ethics after Babel: The languages of morals and their discontents*. Princeton: Princeton University Press.

Sumner, L.W. 1996. *Welfare, happiness and ethics*. Oxford: Clarendon Press.

Sztybel, D. 2006. A living will clause for supporters of animal experimentation. *Journal of Applied Philosophy* 23(2): 173–89.

Thomas, K. 1983. *Man and the natural world: A history of the modern sensibility*. New York: Pantheon Books.

Thompson, P. 2008. Current ethical issues in animal biotechnology. *Reproduction, Fertility and Development* 20: 67–73.

Varner, G. 2002. Animals. In *Life science ethics*, ed. G.L. Comstock, 141–68. Ames: Iowa State University Press.

Warren, M.A. 1997. *Moral status: Obligations to persons and other living beings*. Oxford: Oxford University Press.

9 The Religious Factor: Perspectives of World Religions on Animal Biotechnology

HAROLD COWARD

This chapter examines concerns raised by the development and application of modern animal biotechnologies in the major religious traditions of Judaism, Christianity, Islam, Hinduism, and Buddhism. While religion was rarely mentioned by participants in the stakeholder focus groups conducted as part of this study, the ethical and social concerns raised in the groups relate well to those found in the key texts, traditions, and ethical frameworks of these religions.

A chief concern for all the stakeholder groups was the question of the position of animals in relation to humans and the way in which that understanding conditions our approach to animal biotechnology. Stakeholder viewpoints ranged from those that see animals as existing to serve human needs to those stressing the interconnections of humans, animals, and all of life. In terms of the religious traditions, the first viewpoint reminds one of some interpretations of the human – animal relationship in the biblical Genesis text, while the latter position evokes the Hindu and Buddhist approaches.

In the focus groups, it was clear that one's overall perspective on the animal – human relationship strongly conditioned thinking about such things as stewardship, respect for animals, and where one draws the line between what is acceptable and what is not in the application of biotechnology, a distinction referred to by several participants as the problem of 'playing God.' These questions are important not only for the followers of various religions who happen to be involved in biotechnology, whether as scientists, researchers, activists, or regulators, but also for the sincere devotees of religion who daily make decisions about the food they buy and eat. Thus, for large numbers of religious people in Canada and around the world, concerns over genomic science and its applications in animal biotechnology are not just abstract questions for theologians, but practical matters that determine how one lives one's religion in everyday life.

Before proceeding, it is helpful to clarify how religions are understood in this analysis. Within religions and between religions such as Christianity, Islam, or Buddhism, there is a great deal of diversity. The variety of viewpoints is taken into account through the use of dialogue, focus groups, and the study of key texts and thinkers within each religious tradition. When it comes to ethical issues and wider concerns, such as those raised by modern animal biotechnology, religions focus mostly on duties and virtues rooted in their traditions, and they set up models. They emphasize freedom and responsibilities. Because genomic science and its animal biotechnological applications are relatively recent, responses to these developments in many religions are just beginning to be developed. Theologians and lay followers from these religions find that matters such as cloning, transgenics, and the genetic engineering of animals in general raise new questions and call forth new theological answers from their traditions. These developments also provide new challenges to government regulators in areas such as food labelling so that followers of each religious tradition can exercise choice in practising their ethical views – a religious freedom that is affirmed in the Canadian Charter of Rights and Freedoms (1982), section 2, which guarantees that the state will not interfere in the religious beliefs and practices of any religion.

This chapter begins with an examination in section 1 of a key question arising from the stakeholder focus groups, namely, what is the position of animals in relation to humans? The approach here will be to look at religious responses to this question, and in particular to concerns raised by focus group participants over the implications of animal biotechnology for such things as human stewardship responsibilities, respect for nature, and the hierarchy of being. Section 2 moves to a discussion, drawn from the work of contemporary religious scholars, of specific animal biotechnology applications. Examples discussed will include applications that have led to improvements in the nutritional quality and disease resistance of food, to the reduction of negative environmental impacts, and to human health benefits. Section 3 will focus on contexts where both the stakeholder groups and the religions clearly draw the line on the acceptability of animal biotechnology. Examples of applications deemed unacceptable would include those conducted for merely cosmetic purposes or commercial greed (such as the

GloFish – an aquarium fish that is engineered to glow), those where the basic personality or appearance (identity) of an animal is altered, or those performed on animals (e.g., primates) considered too closely related to humans. Finally, section 4 will discuss some of the policy implications identified by stakeholder groups and the religious traditions. These include the need to go beyond health and safety concerns to address the issue of public acceptance of animal biotechnology, and the need for the labelling of food products so that followers of religious traditions can exercise ethical choice in their purchasing.

Position of Animals in Relation to Humans in the World View of Each Religion

In all focus groups, there was recognition that one's presuppositions regarding the relations of humans to animals in the 'chain of being or becoming' conditioned one's ethics and general concerns. Stakeholder views varied within and between focus groups, ranging from those that see all life forms as so strongly interconnected that species boundaries can not really be identified (Scientists, Animal Justice, Health Care Providers and Patient Advocates) to those which view animals as existing for human use and benefit (Scientists). In a more modulated form, the latter anthropocentric position requires that although humans have a privileged position, still they must be good stewards, treat animals with respect, and not cause animals undue suffering unless essential for human health. In the religious traditions, this idea of the interconnectedness of humans and animals pervades the Eastern religions of Hinduism and Buddhism, while the model of human dominance with stewardship responsibilities characterizes the perspectives of the Western religions of Judaism, Christianity, and Islam.

Interconnectedness of Humans, Animals, and the Environment

As discussants in the Health Care Providers and Patient Advocates focus group put it, there is a strong interconnectedness between humans, animals, and the environment that is linked to feelings about the sacredness of life. This is indeed a good statement of the Eastern world view manifested in the Hindu and Buddhist religions (Coward and Goa 2004: 8–12). The Hindu approach to animals is based on the notions of *karma*, *samsara* (rebirth), *ahimsa* (non-violence), and the presence of the divine in all beings (Narayanan 2009). Animals, for Hindus, are human souls in different bodily forms. Eating an animal is thus quasi-cannibalism. Humans are reincarnated; they may have been animals in past lives, and they may be reborn as animals in future lives. Animals have no free choice, but humans do. Animals have to 'burn off' bad *karma* they built up as humans over many lifetimes of making evil choices. Then they can be reborn as humans with free choice and the ability to move up or down 'the ladder of being.' Hindus also follow *ahimsa*, the doctrine of not harming any living creature, animal or human. For them, the divine exists equally in all beings (*Bhagavad Gita* 5: 18). Animals, like humans, are viewed as manifestations of the divine, which leads to a deep sense of unity and respect for all life forms and their interconnectedness in the divine. Thus, early in Hindu history (i.e., before 200 BCE) hospitals for animals were established in India. This reverence for animals was also the view of Mahatma Gandhi and other contemporary Hindu leaders. As a result, millions of Hindus eat no fish, meat, or eggs. Vegetarian practice is the ideal. However, many others do eat chicken and fish but no red meat. Devout Hindus would refuse to kill animals, but some would eat those killed by others.

Given these precepts, some Hindu philosophers reject all research on laboratory animals, while others would allow some research if it produced results that would relieve human suffering. Genetic modification of animals by humans is seen by many Hindus (both lay people and scholars) as an attempt to meddle with the divine order. Such action renders animals impure for use in religious rituals and holy feasts (Narayanan 2009). Although genetic and biotechnological interventions upon animals might be acceptable in other than religious ritual settings, the human motivation for engaging in animal biotechnology is always suspect. As my Hindu colleague Dr Rambachan puts it,

> Genetically modifying animals presumes a right of human beings to exercise power over other life forms with no concern for their own integrity . . . As a Hindu I want to examine the human motives . . . Too often human greed and the excessive manipulation of wants are the primary motivating factors. The aim of Hindu life is not the maximizing of wants and the manipulation of animals to satisfy these, but the reasonable limitations of wants in the interest of the larger community of beings. This is life in harmony with *dharma* and the spirit advocated in the *Bhagavadgita*. (Rambachan 2004)

Like Hinduism, Buddhism also assumes 'the interconnectedness of all life' (as affirmed by several participants in the Health Care Providers and Patient Advocates focus group). Buddhism adopts the world view of *karma* and rebirth and the resulting ladder of existence on which animals (lower on the ladder) are beings like humans but in a different karmic form. Thus, Buddhists generally believe that they may have been an animal in a past life, and may be reborn as an animal in the future.

In Buddhism, all elements of nature (humans, animals, plants, earth, air, and water) are seen as interrelated and part of a much larger life-force, the Buddha Nature. To do harm or treat with disrespect any part of this entity (e.g., animals) is to harm oneself and all of life. Consequently, the Buddha teaches compassion to all sentient beings. Animals, as sentient beings, are highly respected in Buddhist scripture and teaching. Also, like Hinduism, the ethical teaching of Buddhism stresses *ahimsa* or non-violence towards all living beings. As a general rule, Buddhists refuse to hurt or kill an animal, or to eat meat, though some do choose to eat meat. Tibetan Buddhists eat meat because they are unable to grow crops at the altitudes where they traditionally have lived. But they kill only as many cattle as are really needed. There is no hunting for sport (considered frivolous and hurtful). Buddhists also allow for the possibility of eating meat that wasn't slaughtered specifically for oneself, and many Buddhists eat meat they buy in stores or restaurants. However, devout Buddhists are supposed to live on fruits, vegetables, and grains. A leading Buddhist teacher, Roshi Phillip Kapleau, of the Rochester Zen Center, maintains that vegetarianism is the first precept of Buddhism. To kill or harm an animal, says Kapleau, is to violate the Buddha Nature, the sacred harmony that unites and is manifested in all organisms (Walters and Partness 1992: 84) As was the case for Hindus, eating meat is seen as a kind of cannibalism because of the *samsara* or rebirth presupposition.

Given the Buddhist and Hindu belief in the interconnectedness of humans, animals, and the environment, the use of animals in scientific experimentation is viewed as problematic. If humans engage in genetic modification of animals, from the Hindu-Buddhist perspective this would be acceptable only if there are clear benefits to animals and humans that could not be achieved in any other way. Such activities must be done in a way that does not interfere with the happiness of animals or make them any less able to progress up the ladder of being to rebirth as a human and eventual release (*moksa* or *nirvana*).

In Buddhism, the issue of motivation is key. If animal biotechnology is done for frivolous or purely commercial 'bottom-line' reasons, that is unacceptable. As the Buddhist scholar David Loy puts it, the genetic modification of animals for food may be acceptable if it reduces suffering and if it is done with the intention of bringing about a good result (Loy 2009, 2005). Loy questions, however, whether humans have achieved such a level of awareness about their own motivations. With these stringent requirements in place, it is hard to imagine how any use of animals for food or for biotechnology can be acceptable to Buddhists.

Human Dominance over Animals but with Stewardship Responsibility

Unlike the strong interconnectedness perspective of the Eastern religions, the Western religions of Judaism, Christianity, and Islam see the human – animal relationship as one of animals having been created by God to serve human needs, but with humans having a stewardship responsibility in their relationship with animals. This viewpoint was strongly present in the Agricultural Producers focus group, with members of the Health Researchers focus group also emphasizing the need for human stewardship and respect for animals. As the biblical book of Genesis presents it, humans are created with priority over animals, which are there to meet human needs. While all parts of God's creation are inherently valuable, there is a clear hierarchy of being, with humans at the top and animals lower down. At the same time, both animals and humans are seen to be parts of God's creation, all of which God blesses and sees as good. Therefore, humans in their stewardship responsibility are to minimize cruelty to animals – thus the *kosher* (Judaism) and *halal* (Islam) rules, which are intended to ensure humane slaughtering of animals.

Judaism has several moral and legal rules regarding animals. In the first place, Jews see animals as a part of God's creation for which humans have responsibility; therefore, cruelty to animals is not allowed. The Talmud says that animals should be fed before humans, not restrained unnaturally, and not worked on the Sabbath. The medieval scholars all seem to agree that humans have a responsibility to help animals in distress, for example, by helping a horse to pull a heavily loaded cart up a steep hill.

Second, Jewish law prohibits the cross-breeding of species. The prohibition is especially concerned with practices involving the sexual act between different species; some forms of grafting are permitted, however, where there is no significant change in the appearance of the resulting product. This leads to a third precept, that humans must not interfere with the natural order in ways that change the fundamental appearance of plants or animals. Another aspect of Jewish ethics is the core value of saving life (*pekuach nefesh*) and healing the brokenness of the world (*tinkum olam*) (Zoloth 2009). Using animals in scientific experimentation, for example, is usually permitted by Jewish law (1) when benefit may accrue to the public at large, and (2) because the elimination of the pain and suffering of humans takes precedence over considerations of animal pain. A key principle here is the element of necessity. As one authority states it, 'In any situation in which there exists a human need that cannot be otherwise satisfied, it is not improper to cause discomfort to animals' (Bleich 1986: 86). A final precept is that the Talmud does not list what is permitted, but rather what is forbidden (e.g., the cross-breeding of animals). But when such an activity is done by a non-Jew, a Jew may benefit from it (e.g., by eating broccoli or riding a mule).

Animal biotechnology is now a hot topic among Jewish thinkers (Bleich 2004). An overriding principle from the book of Genesis is that God has made humans as collaborators in completing the work of creation to reach its destined end and spare humans travail. But there is tension between the role of humans in completing the process of creation and the biblical prohibitions against certain forms of interference in the natural order (e.g., the mixing of species). A similar view was expressed by a participant in the Animal Justice focus group who maintained that, like humans, animals have a well-developed telos that should not be 'messed with' in animal biotechnology. That said, Laurie Zoloth, an Orthodox Jewish ethics specialist, notes that most Conservative and Orthodox authorities advocate the widest use of new ideas in agricultural biotechnology (Zoloth 2009). Even transgenics and cloning would seem acceptable, since there is no cross-species sexual activity involved, as long as the overall appearance of the animals has not been radically altered. In general, the overall approach of the Jewish world view and its ethics is that anything that benefits people and is not prohibited is encouraged, as long as there are no associated dangers and it does not cause unnecessary suffering to animals. But can we say that transgenics can be done without causing suffering, even unnecessary suffering, to animals?

Islam is similar to Judaism and Christianity in giving privileged status to humans over animals. Like the Torah, the Qur'an forbids cruelty to animals, but it also goes further to suggest that animals have souls and possess some rationality. Further, all animal species are considered to be communities, like human communities (Masri 1987: 5). Mohammed urged his followers to show compassion to animals because they are part of God's family. In the afterlife, he said, one receives rewards in relation to how we treat animals in this life (Regenstein 1991: 251). Islam also teaches that animals possess a psyche; they have a lower-level consciousness than humans, but it is higher than just instinct. These views put one in mind of focus group comments from Health Researchers suggesting that, on the hierarchy of being, some animals, such as primates (e.g., chimpanzees), share human characteristics associated with the brain too fully to be acceptable candidates for biotech interventions. In line with their views of animals as possessing a psyche, Muslims contend that animals communicate with God. However, humans are judged to have spiritual volition and greater freedom of action. Humans are God's vice-regents on earth. As such, humans have stewardship responsibilities. Like Judaism, Islam condemns frivolous uses of animals for sport, or cosmetics research, or the killing of animals for floor coverings. Animals can be killed only when needed for food, and then only in a ritual way (*halal*) that is meant to minimize suffering. Like Judaism, in Islamic law (*shari'a*) anything not forbidden is permitted.

When questions of genetic modification in agriculture arise, many Muslim scholars go back to an encounter Mohammed had with workers in the field grafting different species of date-palm seedlings. Although at first Mohammed suggested it would be better not to graft, he later limited his authority to moral matters that impact on one's salvation. When it came to practical agricultural concerns, Mohammed endorsed experience and expert opinion. In other words, ethical issues such as the genetic modification of animals that are tied to secular or worldly pursuits and rely on scientific knowledge are to be decided on the basis of their scientific and practical merits (Moosa 2009). Medical research using animals is allowed if laboratory animals are not caused pain. Scientific experiments upon animals are ethical or unethical according to the intentions of those who perform them. All basic and applied research must be shown to be required by human necessity, and the pain upon the animals involved must be minimized (Masri 1988: 192–4). Therefore, medical research would be allowed if the research is necessary, if the motives of the scientists are 'good,' and if the animals are not caused unnecessary pain.

Moving to Christianity, the mainstream attitude in Christianity until recently was that animals are created by God for human use. As a participant in the Animal Justice focus group put it, the book of Genesis teaches that humans have dominion over animals and are empowered to exploit animals to their own advantage however they see fit. Unlike Muslims, Christians do not view animals as having an immortal soul. Christian views on these matters were influenced by the Greeks. In particular, Aristotle exerted influence over Augustine and Aquinas. Aristotle argued that nature made animals and plants for the sake of humans. Augustine followed suit, saying that animals and animal suffering are here for the physical and spiritual benefit of humans. Aquinas agreed, claiming that animals have no reason and no immortal souls. Luther likewise limits rationality to humans and further emphasizes the power of human 'dominion' (Linzsey and Yamoto 1994: 65; Yarri 2005: 107).

This view is now being questioned by many Christian scholars, however, as a misreading of the Bible. Through the ages, there have been minority voices who have been advocates for animals, for instance, Francis of Assisi (Yarri 2005: 107). In her recent reassessment, Donna Yarri argues that human dominion over animals should be understood as benevolent stewardship rather than autocratic despotism (Yarri 2005: 132; Grant 1999). Many Christians now view animals and humans together as parts of God's creation – all of which God blesses as good and inherently valuable. This view is validated in the first chapter of Genesis, in which it appears that humans and animals live together harmoniously as vegetarians, and in the Old Testament example of an agricultural society in which domesticated animals are treated with respect and compassion. More recently, a new generation of Christian environmentalists has come to see humans as part of an ecosystem in which humans and animals are an interdependent part of nature – a nature created by God (see especially Wirzba, 2003). The idea is that animals are suffused with God's Spirit (Nash 1991: 117–21; Reuther 1992: 247ff.; Cobb 1998: 173–80).

Regarding the use of animals in science, Andrew Linzey, taking into account the above theological discussions, offers the following principles. Animals are not instrumental to human ends. Animals are not laboratory tools. Because animals are part of God's creation and interdependent with humans, the motivation behind our use of animals in science, agriculture, or as food must be carefully analysed (as the Buddhists maintain). Animals, like humans, are valuable *in themselves* by virtue of their creation by God. As stewards of creation, humans are accountable to God for how they use animals. Such uses must not be for human ends only, but for the good of the whole interdependent creation (Linzey 1986; 1994: 143–8). In the teaching and life of Jesus, we find a compassion for animals and their pain, and in the Holy Spirit, a hope that as the world struggles towards a New Birth, animals, humans, and all of creation may regain their original state of peaceful coexistence (Romans 8:18–39; Wirzba 2003).

Analysis of Animal Biotechnology Applications

Following the preceding outline of the world views of the major religions towards animals, this section moves to an examination of the implications of these values and precepts for specific animal biotechnology applications. The applications discussed are grouped into four major categories: those aimed at improving the nutritional quality and disease resistance of food and the efficiency of food production; those used to mitigate against negative environmental impacts; those that benefit human health; and those that involve the industrial manufacturing of animals through cloning and transgenics.

Applications to Improve Nutritional Quality, Disease Resistance and the Economic Efficiency of Food Production

A motivation expressed by many of those involved in genomic science and animal biotechnology is that these advancements will benefit the poor and all of humanity by increasing the quality of food and the efficiency of global food production. Participants in the Scientists focus group said that they are involved in animal biotechnology to achieve the greatest good for the greatest number. However, in the Animal Justice focus group, participants expressed the view that in our concern to help people we must not cause pain and suffering to animals. The Jewish religious tradition, with its twin values of *pekuach nefesh* and *tinkum olam* (saving human life and healing the brokenness of the world), supports applications of animal biotechnology that achieve these humanitarian goals, so long as the main motivation involved is not economic greed (Zoloth 2009). For Buddhism, too, the major worry seems to be over the motivation involved (Loy 2009). As Narayanan notes, many Hindus may welcome genetically modified foods (aside from their being proscribed from use in religious rituals or on holy days), for example, chickens that produce more nutritious eggs, so long as there are no health hazards (Narayanan 2009). Islamic scholars would follow Mohammed's example in leaving practical agricultural matters to be decided on the basis of their scientific and practical merits (Moosa 2009). Thus, applications to increase nutritional quality and the economic efficiency of food production could be embraced, as long as the biotechnology in question did not increase animal suffering. If biotechnology applications helped to increase the disease resistance of animals being raised for food, then this would reduce animal suffering and be judged a good thing.

Christians, as noted in section 1, focus on the stewardship principle. One study in particular, *Engineering Genesis*, has examined Christian concerns in relation to the genetic engineering of animals (Bruce and Bruce 1999). The book's key question is how far are humans justified in intervening in the lives of animals for our benefit? While cruelty towards animals is clearly not acceptable, the use of biotechnology to increase milk production or to produce a therapeutic protein in milk is considered ethically acceptable. But respect for animals requires that they be seen as more than just supermarket commodities or generators of bigger profits for producers and retailers. The authors expressed an additional worry that the introduction of animal biotechnology will further foster large-scale agribusiness approaches globally that will force small farmers out of business. Similar concerns were raised by the World Council of Churches in its 2006 report on genetics and agriculture (WCC 2006). Its worry is that the introduction of GM animals in agricultural will reduce biodiversity and result in the loss of the cultures and the traditional knowledge of indigenous peoples and small farmers in developing countries. From this perspective, there is real danger that agricultural biotechnologies, as used by the market economy, may actually further problems of injustice and violence for the world's poor (ibid.: 32, 72). Most adherents to all religions would agree that concerns such as these, along with worries over causing pain and suffering, must be carefully weighed when the potential benefits of animal biotechnology are being considered.

Reduction of Negative Environmental Impacts

Some stakeholders noted that humans have the capacity to control a lot of what happens in our world, for good or evil. Whereas consumer demand for products such as hamburgers, which drives the clear-cutting of Amazon forests in order to produce more cattle for beef, leads to a bad result (increased global warming), the creation of the Enviropig, engineered to have less phosphorous in its manure and be less destructive to the environment, is an example of a technology designed to produce a good result (Golovan et al. 2001). The Eastern religions of Hinduism and Buddhism, with their focus on the strong interconnectedness of humans with nature, would agree. The Western religions of Judaism, Islam, and Christianity, with their stewardship ethic, would also take a favourable view of applications that help to reduce negative environmental impacts. Another example would be the engineering of trout to have a biomarker chip that will detect pollution in streams so that such human-generated pollution can be better detected and regulated (Koop et al. 2008). A further use of such engineered trout will be to more effectively test streams to see if their water is safe for human consumption, thus avoiding health risks – an application that would likely find support in all religious traditions.

Human Health Applications

As a whole, focus group participants identified a number of applications where animal biotechnology is useful in the treatment and prevention of disease. Some examples include the use of animals in pharmaceutical research, the engineering of animals to develop specific diseases for therapeutic research, and the transplantation of animal organs into humans. While all of the religious traditions ultimately give priority to humans over animals when it comes to matters of health, there are still some guiding values and conditions that prevail in the evaluation of these applications. For Christians, whereas the *Engineering Genesis* study found the engineering of dairy cows to produce medically useful proteins to be an ethically acceptable use of biotechnology, some scholars took issue with examples such as the oncomouse, in which animals are engineered to exhibit a specific human disease (in this case cancer) for the purposes of therapeutic research. The authors claimed this would be ethically unacceptable no matter how compelling the reason for doing it. Instead, we should search out alternative methods for doing such research (Bruce and Bruce 1999: 131–40).

Hindus and Buddhists would seem generally to agree with this Christian assessment that the engineering of animals to develop specific diseases is ethically unacceptable. The doctrine of non-violence and the idea of the presence of the divine in all animals would result in the rejection of such applications. However, if animals were not caused to suffer or their original nature was not altered, then their use in the production of pharmaceuticals to help humans would seem to be acceptable. In Islam and Judaism, the general rule that anything not forbidden is permitted would seem to support the scientific use of animals to create therapies for human illnesses, so long as the key element of necessity is present. The test here is whether any other way exists for this human need to be satisfied; if not, then it is not considered improper to cause discomfort to animals (Bleich 1986: 86–9).

Another practice that has become relatively common is the transplantation of organs from animals to humans for health reasons (the most common being the use of pig heart valves in humans). Genetic technology is used to modify the DNA of animals so that their organs are more acceptable to humans and will not be rejected following transplantation. A worry here is that the practice of transplanting animal organs such as hearts, lungs, or kidneys into humans will lead humans to view animals merely as 'spare-parts factories.' This concern was raised in several of the focus groups. Members of the Animal Justice focus group maintained that humans are not more important than animals; most did not support the use of animals as organ donors for humans. Some participants in the Scientists focus group argued that life at the genomic level is so tightly integrated that there is little to differentiate human-to-human from animal-to-human organ transplantation. But others in the same group did make a distinction between the organs involved, arguing that heart or kidney transplantation would be acceptable, but the transplantation of brain tissue would not, because the brain is too closely associated with the personality or identity of the animal or person.

The Christian tradition, as reflected in the *Engineering Genesis* study, sees the use of animals as donors of organs to be an extension of our God-given relationship with animals that would radically change the way animals are perceived. To take a heart from an animal would be a violation of our stewardship responsibility that should only be done in cases of extreme human need and after all other methods, including steps to prevent heart disease through exercise and diet, have been tried (Bruce and Bruce 1999: 131–40). Other religions would seem to agree with this view that transplantation is a major step that must be justified by human necessity. As yet, none of the religions seem to have dealt with the concern raised by the Scientists focus group over the possibility of the transplanting of brain tissue. One guesses, however, that adherents to all the religions might well agree with the view of several speakers in the Regulators and the Alternative Agriculture focus groups who thought that transplants involving brain tissue would be ethically unacceptable and that to do so would be to 'play God' with the nature of an animal.

The Industrial Manufacture of Animals – Cloning and Transgenics

Regarding cloning, some participants in the Agricultural Producers focus group rejected the use of cloning for humans, but found it acceptable as a method to effect improvements in animals. Many in the Scientists and Health Researchers focus groups also rejected the cloning of animals when done for vanity or merely commercial reasons. Frivolous uses such as the cloning of one's pet dog were widely rejected. Among Christian theologians, cloning has raised a number of questions and problems. Some have argued that creating animals on-demand goes against God's plan for biodiversity as set forth in the Bible. It is seen as an act of hubris and irresponsibility, and since hubris is the greatest sin, the cloning of animals is the greatest wrong. But the Church of Scotland has given approval to some small-scale cloning work – for example, for the creation of a small number of GM cattle lines for small-scale medical applications. If the reason for cloning is motivated by profit, convenience, or the demands of human preference, then it is unacceptable as it removes from the animal its 'freedom to be itself' (Bruce and Bruce 1999: 142).

Hinduism and Buddhism, with their stress on human motivation as the determining factor, would seem to be generally in agreement with this Christian assessment. In Judaism, where the major restrictions are against species mixing through cross-breeding via sexual intercourse, cloning seems acceptable since the modification takes place within a single species (Zoloth 2009).

In Islam, the issue of cloning has been examined by three academies of Islamic law (which met in Morocco, Saudi Arabia, and India). A key concern for them was to determine whether humans are attempting to take on the power of creation through cloning, a power which belongs to Allah alone ('For Allah alone is the creator of all things' [Qur'an 39.62]). The Islamic principle here is that science should not create things (Allah's realm), but should rather seek to make understandable the facts of Allah's creation (Qasmi 2003: 9). These scholars see cloning as a miracle made possible by Allah. If successful, it must be because it has the consent of Allah. None of the elements in cloning are human-made – all were made by Allah. So there is no change in the birth order of Allah's creation. The only difference is in how fertilization takes place. Cloning, then, is still an act of Allah, who is the creator of all things, and, thus, is acceptable (ibid.). At this point, Islamic scholars accept cloning for animals, but not for humans, because it would create chaos. They say that research in the field of cloning should be restricted to uses that benefit the world and do not cause undue suffering to the animals involved. While this Muslim position would seem open to the use of animal cloning for purposes such as feeding the world's hungry, it would clearly reject applications that are unconnected with real human need.

In addition to cloning, industrial agriculture uses the process of transgenesis to move a gene that expresses a desirable trait from the same species or another species into the genome of an animal that will then manifest that desirable trait. Resulting animals may be engineered to grow larger and/or more quickly (e.g., transgenic salmon), be less damaging to the environment (e.g., the Enviropig), or be disease resistant.

Among Jewish scholars, there appears to be considerable support for the transgenic modification of animals, since it does not appear to be in violation of the prohibition against cross-breeding (i.e., it does not entail a sexual act between members of different species), and the process seems to involve an approach in which the 'grafted element' (the moved gene) takes on the identity of the species into which it is being grafted, so that there is no significant change of appearance. The *halakhic* (Jewish law) issue at stake is the identity of the resulting genetically engineered entity, which depends in large part on its physical appearance. Although scholars admit there is still ongoing debate, the consensus seems to be that the status of a cow, for example, that has been modified by genes derived from a pig, is still a cow as long as its general appearance is not changed. In effect, the identity of the 'grafted' pig gene becomes submerged in the identity of the animal (in this case, the cow) into which it has been placed. In discussions regarding genetically engineered poultry, the conclusion is that such chickens are *kosher* provided they exhibit the physical criteria of an identifiable species of *kosher* fowl – in other words, that they still look like chickens. Further, even when an animal has received genes from a non-*kosher* animal, it is permitted as food as long as there is no manifestation of the non-*kosher* gene donor. Given this argument, Jews would have no problem eating transgenic salmon.

In Islamic law, as seen in the discussion on cloning above, the debate over transgenic animals rests on the question of whether humans have taken on the power of creation through GM. From this perspective, it would seem that transgenics are acceptable, since none of the elements (e.g., the genes) used in transgenics are human-made – Allah has created them all – and since no change occurs in the birth of the animal or in its natural stages of creation as given by Allah. For Islam, then, according to these scholars, transgenics, like cloning, can neither be called 'creation' nor even a partnership in creating, and is therefore judged to be acceptable. However, the production of transgenic animals must also be shown to be in the best interests of human society (not a cause of chaos or disturbance) and must not cause harm to animals (Qasmi 2003).

Christianity seems to take a more guarded approach to transgenic animals than either Judaism or Islam. Andrew Linzey, a professor of theology and animal ethics at the University of Oxford, argues that animals, like humans, are valuable *in themselves* by virtue of their creation by God (Linzey 1986). As co-creators or stewards of God's creation, humans are accountable to God for the ways in which they use animals. Such uses must not be for human ends only but for the good of all creation. From this perspective, the transgenic modification of animals goes against the God-given natural biodiversity of life. The presumption that humans know what is optimum for selection from the vast diversity and complexity of traits in an animal is an act of hubris (this critique would seem to also apply to ordinary selective breeding). Therefore, the use of transgenics in routine animal production to sidestep normal breeding methods on the grounds of economics or convenience is not acceptable.

Buddhism, in its analysis of transgenic applications, also focuses on the motivation involved. According to the Buddhist scholar David Loy, transgenic animals are not good or bad in and of themselves; it is the human motivation in developing and using them that matters. The Buddhist understanding of *karma* is that actions motivated by negative intentions tend to bring about bad consequences, while actions motivated by good intentions tend to bring good results. If our eagerness to develop and use transgenic animals is motivated by generosity, loving kindness, and wisdom, says Loy, this technology is likely to bring about good results. If, however, we are motivated by greed, ill will, and delusion or ignorance, then we should expect this new technology to increase our suffering and frustration (*dukkha*) rather than reduce it (Loy 2005: 4). This Buddhist approach does not imply that any GM technology is bad in itself. Rather, it is our problematic and confused motivations that tend to lead to negative consequences. Loy offers a Buddhist rule of thumb: 'Is our interest in developing transgenic animals due to our greed or ill will; and . . . can we become clear about why we are doing this? Among other things this means: do we clearly understand how this will reduce *dukkha* [the suffering of humans and animals], and what its other effects will be?' (ibid.: 7). Loy doubts that we have reached such clarity of intention and understanding in our current industrial-agricultural biotechnology.

Another Buddhist scholar, Ron Epstein, focuses on the non-violence principle of *ahimsa* and on the Buddhist respect for all sentient beings (e.g., those with a central nervous system that enables them to feel pain) and their potential for realizing enlightenment. Buddhism, says Epstein, condemns any instrumental use of animals by scientists or anyone. Animals, like humans, are sentient beings and thus must not be treated as objects without regard for their own aspirations. Says Epstein, 'The Buddhist approach to genetic engineering begins with analyzing its effects on life, how it creates or alleviates suffering, and how it aids or cripples the effort of sentient beings [including animals] to realize their potential for enlightenment' (Epstein 2001: 40). He suggests that the development and use of transgenic animals raises ethical questions about the right of humans to alter sentient and non-sentient life on earth (47). Hindus express many of the same worries outlined in this Buddhist analysis.

Where Religions Draw the Line

As the religions consider the issues raised by genomics, genetics, and applications to animal biotechnology, places where they would 'draw the line' are beginning to emerge. While Muslim scholars accept the cloning of animals, the cloning of humans is completely rejected as a cause of chaos in society (Qasmi 2003: 52). Also, any frivolous application or one that would alter the natural identity of an animal is rejected as a human usurping of Allah's role. In both science and biotechnology, all use of animals must be shown to be required by human necessity and to minimize pain for the animals involved (Masri 1988: 192). In Judaism, the Talmud and other authorities are clear that animals are to be fed before humans eat and are not to be worked on the Sabbath (when they must be free to roam the fields). This last requirement would seem to run strongly counter to modern factory-farming practices. According to one authority, the crowded, confined, and inhumane ways in which food animals, such as chickens, are farmed makes it questionable whether or not they can be regarded as *kosher* regardless of how they are slaughtered (Regenstein 1991: 194).

However, here the overriding ethical principle for Judaism is that care and kindness to animals is for the higher purpose of humanizing humans (in their relation with each other) rather than primarily out of concern for the animals. What the Talmud specifically rules out is the cross-breeding of animals. But, as discussed above, both cloning and transgenesis in animals have been found by Jewish scholars not to violate the cross-breeding prohibition or to significantly alter the natural identity of animals (e.g., the 'cowness' of cows or the 'chickenness' of chickens).

In Christianity, earlier thinkers such as Augustine, Aquinas, and Luther all emphasized the principle of human dominion over animals, in which animals are seen to exist only for humans' physical and spiritual benefit. Aquinas allows that cruelty to animals is sinful, but is mainly concerned that cruelty to animals may lead to cruelty towards humans. Due to the theological shift taking place with the advent of Christian environmental ethics, however, humans are now seen to be part of rather than separate from nature. Historically, this may have its roots in St Francis's love for animals. However, Albert Schweitzer starts the modern shift with his extension of Christian love to include 'reverence for all of life' and the requirement that humans, if they cannot refrain from killing animals, must at least be ecologically respectful and just in such killing (Nash 1991: 117–21). Rosemary Reuther notes that creation-centred Christian theologians and philosophers such as Norman Wirzba, Matthew Fox, Teilhard de Chardin, and Alfred North Whitehead offer accounts that overcome the human/nature dichotomy as well as the separation of nature from God (Reuther 1992). The American Methodist theologian John Cobb, Jr, sees God as sacramentally or even incarnationally present in all of nature. He writes, 'To think of all . . . living things as embodying Christ must give us pause. A creature in whom we see Christ cannot be only a commodity to be treated for our gain or casual pleasure' (Cobb 1998: 172). If all are in Christ, observes Cobb, then in some way our treatment of animals is a reflection of how we treat Christ. Such a view clearly rules out any frivolous or instrumental use of animals. It also brings the Christian world view with regard to animals very close to that of Hindus and Buddhists. Cobb concludes that this realization does not mean that Christians will suddenly be able to stop harming animals. But the recognition that, like us, animals are in Christ, will lead humans to wrestle with problems related to their suffering. For Cobb, any application in science or animal biotechnology that causes suffering is ruled out (ibid.: 178).

This Christian position is a serious challenge to the routine use of animals in scientific experimentation. It would seem to rule out, for example, changing the genetic structure of a mouse to ensure it will develop cancer and so be sacrificed for ends that are narrowly human (Bruce and Bruce 1999). It also finds serious problems with the cloning of animals for agricultural food purposes. Linzey extends this objection to the patenting of animals developed via biotechnology. Although animals have always been used in farming, says Linzey, what is new is the use of legal and technological means of subjugating animals so that they become the exclusive property of humans (Linzey 1986; 1994).

Donald and Ann Bruce, in *Engineering Genesis*, raise a further objection, suggesting that experiments in which nuclei for cells are taken from various animals such as pigs, sheep, and monkeys and introduced into denucleated cattle cells constitute the mixing of species in a gross way, going well beyond simply changing one or two genes (Bruce and Bruce 1999). In commenting on the evaluation of GM, Bruce and Bruce observe that, in all discussions regarding GM interventions in animals, a basic opposition of ethical goods arises between what is an appropriate use of animals and the justifiable human benefit arising. They conclude that unless the extreme position of rejecting all animal use by humans is adopted, careful distinctions over acceptable uses of animal biotechnology will have to be continually made. In making these distinctions, Christians must keep in mind the position of animals as fellow beings 'in Christ,' as well as the interdependence of creation, which humans have the duty and the responsibility, as co-creators with God, to protect.

In contrast with the one-life orientations of Judaism, Christianity, and Islam, the presuppositions of *karma* and rebirth lead Hindus and Buddhists to see the question of where to draw the line in a quite different light. Since animals may have been humans in past lives, and will at some point be reborn as humans in the future, the use of animals in science or agriculture should be viewed with the same ethical restrictions one would use if they were human now. In scientific experiments, this means animals deserve the same health, safety, and intrinsic-value considerations one would give to humans. In agriculture, the implication is that while animals can aid humans by pulling ploughs, for example, or providing dairy products, animals themselves should not be used for food; hence the vegetarian ideal. And whereas the Western religions have agreed to the sacrifice of animals in a laboratory environment for human health benefits, the Eastern religions are much more reluctant to accept such human treatment of animals. Not only is it seen as tantamount to engaging in the imprisonment and killing of beings with souls, but such treatment of animals will, in the Eastern view, also result in suffering in future lives for all of the humans involved (Chapple 1986). The resultant suffering will not only be visited upon individual scientists, but upon the whole society which allowed animals to be used in processes where their intrinsic nature as future human beings is ignored.

Evidence of these consequences can already been seen in the negative aspects of science which now plague the world, such as death and disability from adverse drug reactions (e.g., the thalidomide tragedy), increased militarization, and ecological destruction from unsustainable agricultural practices (human violence upon the earth, air, and water).

Buddhists express similar worries about the future results of genetic experiments upon animals. While such activities may help to relieve human suffering in this life, from the Hindu/Buddhist long-term perspective of being reborn over and over until one reaches *nirvana* or enlightenment, such efforts pale into insignificance and are not worth the added suffering (*dukkha*) they bring to the scientists, animals, and the societies involved. From the Hindu and Buddhist perspectives, the living of a disciplined life, incorporating proper diet and exercise and the moderate use of medicines (preferably herbally based) when necessary, is encouraged. When death comes, it is to be accepted gracefully, without attachment (as seen, for example, in attempts to artificially prolong life with transplants of animal organs) or fear. In these traditions, such a 'good death' will result in a good rebirth, through which the ultimate goal of enlightenment (*moksa* or *nirvana*) may be realized. But until all beings, both human and animal, reach enlightenment, the practical problem remains of where to draw the line on the use of animals in science or agricultural biotechnology.

Before passing judgment, Buddhism requires that three factors be considered: the intentions of the act, the means used to do it, and its consequences. To illustrate this approach, the Buddha told a story in which an evil man on board a ship threatens to kill all 500 passengers. The captain has to decide what to do. After considering the intention and the consequences of performing an act of violence, he kills the man, thus saving the 500 on board. Similarly, says the Buddhist scholar Christopher Chapple, in the case of whether to use animals in scientific research, the three considerations of intentions, means, and consequences would need to be considered in each situation. Many current uses of animals would be deemed unnecessary. Only in exceptional cases would the intention be deemed acceptable – such as the testing of a vaccine desperately needed to prevent an epidemic. The means employed would have to ensure that pain to the animals is minimized, and the consequences considered – will lives in fact be saved? Will unintended reactions such as genetic damage, increased cancer risks, or the loss of biodiversity also occur (Chapple 1986)? Such considerations, when used with care, would constitute a reasonable approach to evaluating the use of animals in biotechnology applications for some Hindus and Buddhists. Others, however, would reject altogether any attempt to justify the use of animal biotechnology.

Having reviewed how and where various religious traditions draw the line, it is helpful to consider as well some areas of agreement. All religions would seem to share a common conviction that frivolous applications of animal biotechnology, such as the GloFish, cosmetic research, or the cloning of pets, are seriously questionable from a moral point of view. There is also a common focus on motivation – especially in the Eastern traditions of Hinduism and Buddhism. If the application is meeting a real human need it may be seen as acceptable. However, if it primarily reflects individual or corporate greed or a scientific drive to be first (hubris, vanity), then it is not viewed positively by any religion – nor indeed by the majority of the stakeholders interviewed in this study. Finally, the concern that the telos or species integrity of animals may be challenged by some kinds of genetic modifications was raised in the stakeholder focus groups and by the theologians of many religions.

Engaging in the kind of acts described above with animals, which are divinely created, generates a sense of abhorrence among lay people and a view that humans are overstepping their stewardship limits when they change the essential nature and identity of an animal. The religions are just beginning their analysis of genetic applications and have yet to compare them with other alternatives that would, for example, be just as effective in meeting environmental challenges. For example, Tariq Ramadan, arguably the leading scholar of Islam in the West, says that reflection about respecting the environment or about how animals should be treated is virtually non-existent in Islam (Ramadan 2009: 233).

Policy and Regulatory Concerns from the Religions' Perspectives

Like NGOs, animal rights groups, ethics committees, and various secular publics, members of religious traditions engage a wide segment of civil society and have distinct ethical views about animal biotechnology that deserve to be included in public policy and regulatory decision making. In considering the acceptability of biotechnology, religious traditions address a broader spectrum of issues than just scientific and regulatory ones. Religions tend to focus on moral issues, such as the place of animals in the natural order, which the formal discourses of law and science typically rule out of bounds. Religious perspectives on the relationship of humans and animals depend on a number of presuppositions concerning the divine order of creation, the nature (i.e., soul, rationality) attributed to animals, and the manifestation of the divine in and through them. Consequently, the genetic modification of animals, whether for research or commercial purposes, raises ethical concerns that are very important to followers of these traditions.

Religious views and beliefs about animals are typically expressed in the form of dietary restrictions. In a country such as Canada, with its multicultural and religious diversity (representing all the religions discussed here), there is strong interest in clear and detailed labelling of commercial food items sufficient to give consumers the ability to select food that does not violate their religion's food proscriptions. For example, Hindus and Buddhists practising the vegetarian ideal of their traditions must be able to be confident that what they are purchasing and eating contains no animal materials. The same is true for secular vegetarians. Christians who hold theological convictions about the genetic modification of animals (for any one of the reasons discussed earlier) may wish to avoid GM foods in any form. Consequently, clear labelling seems especially important in a country such as Canada, where freedom of religion is specified in the Canadian Charter of Rights. As one member of the Health Researchers focus group (who self-identified as a Christian) put it, 'My church creeds talk about respect for nature. That pervades what I do in my work . . . and I think it pervades policies for the protection of human health that the Canadian government implements.'

In a recent study of the acceptability of genetically modified foods for members of religious traditions, Conrad Brunk, Nola Ries, and Leslie Rodgers gave special attention to the regulatory implications of religious dietary practices (Brunk et al. 2009). Responding to the views expressed by groups of lay people from the major religious traditions, they drew the following conclusions:

- Nearly all religions have beliefs that place limits on the production, preparation, or consumption of food. These practices will manifest themselves in consumer acceptance of new food technologies.
- For these religions, DNA is ontologically and ethically significant. Thus, transgenes from animals considered impure or inappropriate for consumption may constitute a 'contamination' of foods into which they are transferred, and are likely to be met with rejection by consumers.
- Religious adherents need information not only as to whether a product contains genetically modified organisms, but also about the source of any transgenic material.

Brunk et al. conclude that it is 'incumbent upon regulators of food technology to establish mechanisms that require public access to the information about the origin of any transgenes in genetically modified products.' The dietary concerns of these religious communities, says Brunk, fall within the fundamental rights of religious and moral conscience to which a liberal democratic society should ascribe special weight and respect.

Policy and regulatory concerns also focus on *animal welfare* issues. As Sheila Jasanoff and colleagues of the Harvard School of Government point out, animal welfare issues, including but going beyond pain and suffering, require policymakers and the public to take sides between different philosophical and (in this case) religious theories – 'most notably, between utilitarianism and approaches that may absolutely protect some aspects of an animal's existence against human intervention' (Jasanoff et al. 2006: 13). Jasanoff et al. also raise the question of the need for policy on what she calls 'unnatural creation,' referring to the creation of 'chimeras.' Chimeras are organisms that contain genetic material from more than one species (for example, human embryos containing non-human genes and vice versa). Laboratory-produced genetic crosses, says Jasanoff, are hard to classify and 'raise new ethical questions about the degree of similarity between species (e.g., when does a chimp become human?), the permissible limits of interference with categories established in nature, and the possible unintended consequences of such experiments' (ibid.: 7).

Exactly these questions were raised by participants in our Health Researchers focus group. Among the religions, it seems clear that Muslims especially would reject chimeras as examples of humans taking on the powers of creation, which belong only to Allah. But all the religious traditions express worries over the possibility of genetic interventions that would change an animal's identity, a result they would view as an example of frivolous and sinful human hubris – using our scientific and technological power to do something just because we could, without respecting ethical limits. Such a question was discussed in the Canadian courts when the Supreme Court in 2002 refused to patent the Harvard University genetically engineered oncomouse®. At a time when the boundary between animals and humans is becoming blurred by genetic modification, the court decided that such questions need to be evaluated by parliament (Jasanoff 2006: 13). In such situations, the views of the religious traditions, along with expressions from other publics, would need to be considered as part of the parliament's decision-making process.

References

Bleich, D. 1986. Judaism and animal experimentation. In *Animal sacrifice: Religious perspectives of the use of animals in science*, ed. T. Regan, 183–92. Philadelphia: Temple University Press.

– 2004. *Implications of genetic engineering from a Jewish perspective*. Working paper no. 83. New York: Benjamin N. Cardozo School of Law, Jacob Burns Institute for Advanced Legal Studies.

Bruce, A., and Bruce, D., eds. 1999. *Engineering genesis*. London: Earthscan.

Brunk, C., Ries, N., and Rodgers, L. 2009. Regulatory and innovation implications of religious and ethical sensitivities concerning GM food. In *Acceptable genes? Religious traditions and genetically modified foods*, ed. Conrad Brunk and Harold Coward, 115–34. Albany: State University of New York Press.

Canadian Charter of Rights and Freedoms. 1982. Section 2. http://laws.justice.gc.ca.en.charter.

Chapple, C. 1986. Non-injury to animals: Jaina and Buddhist perspectives. In *Animal sacrifice*, ed. Regan, 213–36.

Cobb, J.B., Jr. 1998. All things in Christ. In *Animals on the agenda: Questions about animals for ethics and theology*, ed. A. Linzey and D. Yamamoto, 173–80. London: Illini Books.

Coward, H., and Goa, D. 2004. *Hearing the divine in India and America*. New York: Columbia University Press.

Epstein, R. 2001. Genetic engineering: A Buddhist assessment. *Religion East and West*, 1 June: 39–47.

Foltz, R.C. 2006. *Animals in Islamic tradition and Muslim cultures*. Oxford: Oneworld.

Golovan, S.P., et al. 2001. Pigs expressing salivary phytase produce low phosphorus manure. *National Biotechnology* 19 (August): 741–5.

Grant, R.M. 1999. *Early Christians and animals*. London: Routledge.

Jasanoff, S., et al. 2006. *Engineering animals: Ethical issues and deliberative institutions*. Prepared for the Pew initiative on food and biotechnology, Michigan State University.

Koop, B., et al. 1986. The place of animals in Creation: A Christian view. In *Animal sacrifice*, ed. Regan, 115–48.

– 1994. *Animal theology*. Chicago: University of Chicago Press.

– 2008. Effects of diesel on survival, growth and gene expression in rainbow trout. *Environmental Science and Technology* 42(7): 2656–62.

Linzey, A. 1998. Introduction to *On the agenda: Questions about animals for ethics and theology*, ed. A. Linzey and D. Yamamoto, xi–xx. London: Illini Books.

Linzey, A., and Yamamoto, D., eds. 1994. *Animals on the agenda: Questions about animals for theology and ethics*. Chicago: University of Chicago Press.

Loy, D. 2005. Remaking the world or remaking ourselves? Buddhist reflections on technology. http://ccbs.ntu.edu.tw/FULLTEXT/JR_MISC/101792.html.

– 2009. The karma of genetically modified food: A Buddhist perspective. In *Acceptable genes?* ed. Brunk and Coward, 179–96.

Masri, Al-Haliz, B.A. 1987. *Islamic concern for animals*. Hants, UK: The Athlone Trust.

– 1988. Animal experimentation: The Muslim viewpoint. In *Animal sacrifice*, ed. Regan, 171–97.

Meyer, G. 2006. *The cloning of farm animals: A European public affair*. Report prepared for the project Cloning in Public by the Centre for Bioethics and Risk Assessment, The Royal Veterinary and Agricultural University, Rolighedsvej, Denmark.

Moosa, E. 2009. Genetically modified foods and Muslim ethics. In *Acceptable genes?* ed. Brunk and Coward, 135–8.

Narayanan, V. 2009. A hundred autumns to flourish: Hindu attitudes to genetically modified foods. In *Acceptable genes?* ed. Brunk and Coward, 159–78.

Nash, J.A. 1991. *Loving nature: Ecological integrity and Christian responsibility.* Nashville: Abingdon Press.

Qasmi, Qazi Mujahidul Islam, ed. 2003. *Cloning in the right of Shariah.* New Delhi: Islamic Fiqh Academy.

Ramadan, T. 2009. *Radical reform: Islamic ethics and liberation.* Oxford: Oxford University Press.

Rambachan, A. 2004. Personal communication, 30 December.

Regenstein, L.G. 1991. *Replenish the earth*, 183–92. New York: Crossroad.

Reuther, R. 1992. *Gaia and God: An ecofeminist theology of earth healing.* New York: HarperCollins.

Waldau, P., and Patton, P., eds. 2006. *A communion of subjects: Animals in religion, science and ethics.* New York: Columbia University Press.

Walters, K.S., and Portness, L., eds. 1992. *Religious vegetarianism: From Hesiod to the Dalai Lama.* Albany: State University of New York Press.

Wirzba, N. 2003. *The paradise of God: Renewing religion in an ecological age.* Oxford: Oxford University Press.

World Council of Churches. 2006. *Transforming life: Genetics, agriculture and human life.* Geneva: World Council of Churches.

Yarri, D. 2005. *The ethics of animal experimentation: A critical analysis and constructive Christian proposal.* Oxford: Oxford University Press.

Zoloth, L. 2009. When you plow the field, your Torah is with you: Genetic modification and GM food in the Jewish tradition(s). In *Acceptable genes?* ed. Brunk and Coward, 81–114.

10 Issues of Governance in Animal Biotechnology

CONRAD G. BRUNK AND SARAH HARTLEY

The previous chapters in this collection reflect a wide variety of viewpoints that are brought to the table in the public discussion around the social, ethical, economic, and environmental values at stake in the development of animal biotechnology. Together they provide a reasonable 'inventory' of the values that are most salient for people in the social roles they play – as professionals, entrepreneurs, regulators, advocates, and concerned citizens.

Among the central concerns around animal biotechnology reflected in these chapters are those related to the proper governance of this new technology. These concerns were raised by the authors of the chapters in their attempt to characterize accurately the implications of the technology for their 'stakeholders.' But they also generated vigorous discussion among the participants in the focus groups who 'represented' in some rough sense these stakeholders. Concerns about the 'governance' of animal biotechnology are not just about its regulation by government regulatory agencies, but also about other political and social tools for controlling what is done to animals, how it is done, and the roles of government, industry, advocacy groups, consumers, and citizens in shaping the development and implementation of the technology. The authors of these chapters and many of the participants in the focus groups that were part of this study felt that aspects of this new technology required careful social scrutiny, and they were not certain that the current mechanisms of governance at work in Canadian society were adequate to deal with it.

When most people think about the governance of a new (or old) technology, they think about government regulatory bodies. In Canada there is a relatively high level of public confidence in the ability of government regulatory agencies to provide protection from the risks to human health and to the natural environment. This confidence was generally reflected in our focus groups (though certainly not by all participants). The assumption is widespread that new products introduced onto the Canadian market have been thoroughly tested with respect to their health and environmental impacts. This assumption naturally reinforces the expectation that the same thing will be true with respect to new animal biotechnologies. Although there is fear among some that genetically modified plants and animals may pose health and environmental risks that are not taken seriously enough by the regulatory system, this fear does not appear to be shared by the majority of the population in either Canada or the United States.

This confidence, however, does not appear to extend to the regulation or governance of the social, ethical, and economic aspects of these technologies. There is general confusion about these aspects; many people assume that government regulatory agencies in Canada and the United States have the power to regulate aspects of new technologies beyond those of health and environmental safety. Some assume that the other ethical, social, and economic matters are also managed by government. Others believe that government agencies simply ignore them, or handle them in favour of industry or other vested interests. Few seem to understand how the mandates of the regulatory agencies limit their jurisdiction or remit to matters of health and environmental risk.

It could be argued that government agencies are to some degree responsible for this public confusion, because, as will be noted below, they sometimes hold public consultations that include stakeholder groups whose major concerns centre around these broader issues. The implications of this, at least in the public mind, are that these issues are within the mandate of the government agencies. It is not altogether surprising, then, that when these agencies 'fail' to take these issues into account when assessing a new technology or product market approval, it can give rise to some cynicism about the responsiveness of government to these issues.

In this chapter, we wish to examine the expectations that emerged for governance of animal biotechnology in the research undertaken for this book. Some of these expectations have been articulated by the authors of the various chapters. However, they were discussed most extensively and intensely in the focus groups we conducted. This was not necessarily because the governance issues were the most 'top of mind' in the focus groups, but because the groups were asked to respond to specific questions about their views on certain applications of the technology, and how these concerns should be handled. As explained earlier in the Introduction, the groups were asked to identify their strongest concerns about the genetic modification or cloning of animals, and were asked as well to give their responses to a series of technologically feasible applications of these techniques.

In the latter half of the chapter, we describe the existing governance frameworks in place in Canada, the United States, and Europe and examine some of the insights these cases provide. Given our experiences with genetically modified plants, conversations with Canadian regulators, and the results from our research, we draw some predictions about how governance of animal biotechnology will unfold in Canada over the coming years and make some suggestions for how it might be improved.

Focus Group Concerns about the Governance of Animal Biotechnology

In the introductory chapter, we summarized the general attitudes expressed in the different stakeholder focus groups about potential applications of animal biotechnology. In addition, all focus groups were asked how they felt society should address or govern their concerns. We noted in the introductory chapter that there was a surprising commonality among the stakeholders represented in these groups around the ethical and other values at stake in animal biotechnology, even though these values were not always prioritized in the same way. However, on the questions of *how* these values ought to be brought to bear on the development and use of genetically modified or cloned animals and of *who* ought to shoulder the responsibility for doing so, there was much wider divergence of opinion. Few, if any, of the stakeholders argued that the development of these technologies ought not to be constrained by the social values they personally felt were important, or even by those values they recognized as shared by many others in the society. We did not hear much broad argument that the constraint of this technology by moral and other social values would place undue constraints upon technological innovation. The concern about constraint on innovation was triggered mostly by the prospect of such governance being exercised by government regulatory agencies. Other options, such as the acceptance or rejection of certain animal biotechnologies in the marketplace, self-regulation by industry, or even legislative prohibition of morally objectionable forms of the technology, did not raise the same level of concern.

The divergence of opinion on the governance of animal biotechnology followed a somewhat predictable pattern, in terms of both the types of governance mechanisms identified and the preferences for these among different stakeholders. For example, participants from the Alternative Agriculture, the Animal Justice, and the Health Care Providers and Patient Advocates focus groups were strongest in the opinion that social values should govern this technology through government regulation. They generally wanted to see public values better incorporated into public policy through both legislative and regulatory means.

Participants in these groups generally believed that there were many mechanisms, such as stakeholder consultations, citizen juries, and surveys, for involving the public in the regulatory process that governments, particularly in North America, could utilize but had failed to do so. They complained that even when government did engage the public, it failed to act on the recommendations it received. Generally these groups wanted to see industry, consumers, and the public at the table for government-led consultations on animal biotechnology policy and regulations. They expressed scepticism about the effectiveness of the mechanisms that government currently invokes. For example, a member of the Alternative Agriculture group expressed the concern about public participation this way:

> So it's really important to have . . . different stakeholders be given a voice on the national level as well as the international level and when those groups are given a voice, not to just dismiss them out of hand. The Royal Society Report that came out six years ago on biotechnology; and how many of their recommendations have been implemented over the last six years or so? Very few from what I understand. So, it's important to actually have the mechanism, not just to hear lots of different voices, but to make sure that those voices are implemented. (Alternative Agriculture stakeholder V1)

A member of the Health Care Providers and Patient Advocates group offered a similar observation:

> I have a problem with Canada's oversight and governance in a lot of areas . . . In Canada there are all these commissions and studies and oh,

> we'll have to study this and we'll have to study it some more, and then we'll have to study it some more, and then a bill is brought to the House, and it just kind of stalls because nobody seems to want to have the intestinal fortitude to actually take it and run with it and make some decisions about it. (Health Care Provider and Patient Advocate stakeholder V1)

Participants from the Health Researchers focus group generally agreed with the proposition that regulatory policy should reflect public values, but worried that the public was too uninformed and influenced by poor media representations of science and technology. There was general agreement that the best mechanism to incorporate public values was representative and fair consultations that include the public. As one member put it, 'The reason we want the government involved in [these consultations] is presumably because they're responsible to the people. And so the assumption is that somehow that's representative of what the majority of the population wants' (Health Researcher V1). On the other hand, the Agricultural Producers and the Scientists focus group participants generally took a notably different stance with respect to government legislative or regulatory governance of this technology. They generally recognized that some science-based regulation is critical for the commercial development of animal biotechnology products and for consumer confidence, but felt that society is already over-regulated. In their view, decisions about ethical and social aspects of the technology were better left to the market or to self-regulation on the part of the researchers and producers. The Producers group reflected the concern that societal values tend to be in constant debate, and fluctuate over time, introducing far too much uncertainty into the regulatory system. Some members of the Scientists focus group shared these concerns about considering values in public policy, arguing that science shared a 'common truth,' whereas values or ethics were heterogeneous and could change rapidly:

> I get pretty nervous when people start introducing ethical considerations into the regulatory decision-making process, because I just feel that is – it could change with each government . . . You know, you can't exist in society where every eight years or four years all the rules change because the new . . . president has a different moral standard to the previous one, or whatever. And that makes me nervous, to take the science basis of regulations and interject ethics. I mean, who has the same ethics as anybody else?' (Scientist V3)

One of the Agricultural Producers expressed the concern about over-regulation by government, and preference for governance via the marketplace, in this way:

> I think government has got a responsibility to protect human health and environmental health. Leave the rest to us. Get out of our way . . . because if you try to quantify the other stuff, that's called marketing. We read the market, we tailor what we are producing, we'll provide a stream and we'll sell it, and if there's a buyer, as long as you're not risking anything; we've got to find a way to be a less regulated society. We are becoming very highly regulated, we're becoming too European in our approach and it's dampening our ability to respond. (Agricultural Producer V4)

Interestingly, some members of the Producers group pointed out that the export-dependent nature of Canadian agriculture suggested that Canadian social values were not as significant as those that would likely be expressed in the global marketplace. If certain aspects of animal biotechnology created ethical concerns at home, they might not do so abroad, or vice versa. This concern appeared to arise out of the experience of Canadian agricultural producers with GM crops, such as soy and wheat, which did not meet serious opposition at home, but did so in Europe, Africa, Japan, and elsewhere. In fact, Canadian farmers largely opposed the application from a major seed company to have its GM wheat variety approved for the Canadian market, because of fear of the loss of export wheat markets. In this case there was at least a tacit recognition that to leave the question of the largely ethical objection to GM crops to the global marketplace could have produced an economic disaster for Canadian wheat farmers.

Several participants in the Regulators focus group emphasized that the existing regulatory framework did not permit them to consider matters other than the environmental and human health implications of animal biotechnology that could be assessed scientifically. However, participants recognized that animal biotechnology needed to be guided by ethical and other public values, but felt that these should be incorporated into public policy at other points, such as special commissions fashioning legislation to prohibit certain forms of the technology. They also insisted that regulators to some extent already incorporate societal values into the existing regulatory regulations governing animal biotechnology, largely in the way that human health and environmental safety are understood and safety standards implemented. However, they recognized that this still did not permit the incorporation of the full range of ethical and other concerns surrounding the technology.

Several participants in the Regulators group suggested that it is the democratic parliamentary system, not the regulatory bureaucracy, that appropriately allows Canadian policy to reflect societal values on a large scale through legislation. But some of the Regulators contended strongly that the industry needed to exercise more responsibility for providing consumer information on the food products derived from animal biotechnology in order to build consumer confidence in their products (and avoid those products that did not inspire consumer confidence). The implication of this view was clear, even if not made explicit – that information to consumers was best handled in a system of voluntary labelling of animal biotech products rather than a mandatory system, which shifted the responsibility more heavily upon the regulators themselves.

A common issue arose in the focus group discussions, regardless of the position that was taken on whether the governance of the ethical and social value questions raised by animal biotechnology should be exercised by parliament, regulators, industry, or consumers themselves. It was the issue of the availability of information about products of the technology, usually understood as involving whether or how these should be labelled.

Those who were most supportive of strong government roles in shaping and regulating the technology felt that a system of mandatory labelling of cloned or genetically engineered animal products was a critical part of that regulatory process in so far as it contributed to the transparency of the regulatory process by making information about products publicly available, while also empowering consumers to reflect their own values in the market. As one Alternative Agriculture stakeholder (V3) voiced it,

> I want to know a lot. I want to know what all the different sources of DNA, you know, would be really important. You know, if there's something that's a combination of DNA in a number of different sources that would be one example.
>
> [Moderator:] 'And what ways would be best to get that information?'
>
> I think labelling definitely should be required.

The Health Care Providers and Patient Advocates, Alternative Agriculture, and Animal Justice groups articulated this view most strongly. Participants suspected that the public would demand labelling of food products derived from animal biotechnology, but some had concerns about the public's ability to understand labels, for example, the usefulness of simply labelling an animal product as 'GMO' or 'GM-free.' Furthermore, they recognized that labelling animal products as genetically engineered presented serious challenges for an effective system of traceability (of GM animals and food products derived from them).

On the other side, those most opposed to government regulation of the technology – particularly the Agricultural Producers and the Scientists – believed that the ethical and social issues connected to the technology were best handled in the marketplace by consumers expressing their approval or disapproval of a product. Despite the fact that these stakeholders had high confidence in the power of the market to deal with these issues, they reflected the widespread industry suspicion of product labelling as an effective way to communicate to consumers. In discussion of this issue, there was no clarity around how information relative to ethical and social choices might better be communicated to consumers. Yet, when probed on the issue of labelling, this very interesting exchange took place in the Agricultural Producers group:

> You know, as strange as it sounds, and I'm a supporter of a biotechnology, but I think I'd like to know that it's biotechnology that I'm buying, as opposed to regular. (V4)
>
> [Moderator:] 'How would you get that?'
>
> I don't know. (V4)
>
> I have a child who is anaphylactic to peanuts. We've known about checking labels for a long time now. I'm very concerned about labels. I never used to pay attention to labels until we found out we had an anaphylactic child. I would go along with #4, our whole labelling scheme needs a complete revamping for a lot of different reasons. (V3)
>
> [Moderator:] 'What would you like to see on the label of products of animal biotechnology?'
>
> I haven't thought that deep into it, whether I know how to resolve that, but I certainly would like the right to choose and say, because we have to look so carefully at labels now, I can tell you that there are some reasons that you really want to be able to choose. And even if I can't justify it exactly why I want to choose, I do want to choose. I want the ability to choose. (V3)

The Regulators found themselves somewhat caught in the middle on the labelling issue. Whether a labelling regime is purely voluntary or mandatory for GM products, the regulatory agency usually has to set the standards of what can or must be said on a label, and then enforce those standards on the industry. The Regulators generally expected a strong public demand for labelling of GM animal products, because the industry would likely fail, as it had in the case of GM plant products, to inspire the requisite level of public trust in the technology. Overall, they felt that industry had the primary responsibility to educate consumers and the public about the nature of the genetic engineering used in the animals (the process and the product), and to respond to public concerns that might be raised. If this were done properly (it was not clear how, perhaps through advertising), then there would be less demand for labelling of the products, either voluntary or mandatory.

One Regulator put the point this way:

> What I resent as a regulator, I think, is having to deal with [the labelling issue] when really we wouldn't be dealing with the issue necessarily if there was no stigma attached to some of these products, and were the companies to actually market them properly. And they should be the ones front and centre on trying to build public acceptance and making sure that they've got the marketing straight before they develop the products. Then if companies make products consumers want, there should be no problem with distinguishing those products in the marketplace but . . . it's not really government's job to be the marketing arm for a company. (Regulator V2)

A point that came up regularly across nearly all the focus groups was that neither Canadians nor Americans had developed robust democratic processes for dealing with the broad ethical and social issues posed by the implementation of new technologies generally. There was a general view that these societies did not provide effective forums in which there could be exchanges of views among stakeholder groups with different perspectives on new technologies, including industry and government, and that even such forums as were used from time to time were not effective in incorporating widely held public views into government or industry action. Some of the participants were aware of processes such as citizen juries and other forms of stakeholder consultation that they saw more developed in Europe and elsewhere. As one of the Health Care Providers and Patient Advocates put it: 'So I think notions about participatory democracy, we need a lot . . . We need better ways of having a sort of deliberative democracy, because just getting opinions isn't good enough. They have to be informed. I think that's why you asked, "What do you mean by public input?"' (Health Care Providers and Patient Advocates stakeholder V3). There was a widespread assumption across all the focus groups that our system is not adequately structured to provide places where the ethical and other social values raised by this and other technologies can feed effectively into decision making at the level of either public policy or private industrial innovation and production. We turn now to look more closely at how the system works in Canada, the United States, and Europe.

Existing Governance Frameworks for Animal Biotechnology

Canada

At the time of this writing, no biotechnology-derived animals or products have been approved for release into the Canadian environment or for food. The regulatory authorities with a mandate to govern animal biotechnology products have yet to announce a decision on how they will handle the products of animal biotechnology. Until Canadian officials announce legislative or regulatory changes, the existing regulatory framework in place applies to all areas of animal biotechnology. This framework is founded on the 1993 Federal Regulatory Framework for Biotechnology, which is product-based in that it regulates the products and not the process through which they are produced. The system relies on the concept of 'substantial equivalence,' whereby a product is compared to its conventional counterpart, and if judged to be substantially equivalent in terms of its health or environmental impacts, it is considered acceptable for commercial use.

Responsibility for regulating animal biotechnology products is shared between Environment Canada (EC), Health Canada (HC), and the Canadian Food Inspection Agency (CFIA): EC and HC are responsible for environmental safety under the Canadian Environmental Protection Act (CEPA); HC assesses food safety under the Novel Foods Regulations; and the CFIA is responsible for feed safety assessments, under the Feeds Act and Regulations, and animal health risk assessments under the Health of Animals Act. Medical devices, xenotransplantation, therapeutic biologicals, cosmetics, veterinary drugs, and industrial chemicals or biochemicals fall under the Food and Drugs Act administered by HC and enforced by the CFIA.

Although Canadian regulators have yet to make a policy statement on the regulation of animal biotechnology, there is work going on behind the scenes that includes the establishment of inter-agency committees to discuss developments in the field and possible regulatory responses. The regulatory officials we engaged in this study recognized the need to consider public concerns and understand public opinion as part of this process. In an attempt to understand stakeholder (rather than general-public) opinion and concerns, HC (2003) and the CFIA (1998, 2003, 2004) held stakeholder consultations on regulating animal biotechnology as they considered their options in response to emerging animal biotechnology products. The CFIA stakeholder consultations brought together a rather narrow range of stakeholders, dominated by industry and government representatives. Public-interest and consumer groups were absent at the 2004 consultations, despite the fact that their absence had been noted at the 2003 consultations and the CFIA was advised to invite such groups to future consultations. As discussed earlier, our focus group research showed that the Agricultural Producers held a different perspective from other stakeholders, particularly on regulatory issues. This data suggests the information collected through the CFIA's consultations will not be generally reflective of societal views or those of a broad range of stakeholders. In line with the view expressed by our Agricultural Producers focus group, the CFIA's 2004 consultation concluded that regulations should not unnecessarily inhibit the industry, regulations should focus on science, not values (except economic values), and that social and ethical issues should be considered in an outside forum. Self-regulation was considered the preferred approach for the future, because the government will not be able to provide a regulatory oversight of all aspects of biotechnology (CFIA 2004). In addition, the Agricultural Producers thought that a traceability system to track animals produced through animal biotechnology would undermine public trust in the regulatory system by injecting suspicion about regulatory confidence in the safety of genetically modified or cloned animals.

While the Canadian government and its regulatory authorities are considering the policy and regulatory options for governance of animal biotechnology, Canadian industries at the time of this writing have agreed to comply with a voluntary moratorium on releasing these products into the food supply system or the environment. Industry self-regulation of this sort has been a powerful tool in the past. For example, in 1999, McCain Foods announced it would no longer use genetically modified potatoes in its food products in Canada. Although these genetically modified food products had been approved by the CFIA, McCain Foods was responding directly to consumer fears within both its domestic and global markets about environmental and human health risks, and a broader set of concerns related to agricultural justice. By 2002, McDonald's, McCain Foods, Pringles, and Burger King had also rejected the use of GM potatoes in the United States. This response supports the view of those who say that industry does respond to consumer concerns and, arguably, it knows its market. However, it may be risky to leave regulatory decisions to other forms of governance.[1] For example, in the case of GM wheat in Canada, the regulatory framework did not have the legal mandate to withhold approval of a crop variety that represented a potential market collapse for Canadian wheat farmers. The Europeans and Japanese (the primary importers of Canadian wheat) were at the time excluding imported GM wheat, and they threatened to shut out all Canadian wheat imports if GM wheat were approved. Canadian wheat farmers were unable to guarantee pure GM-free wheat in their exports, due to the lack of a system to prevent the mixture of GM with non-GM grain, either in the field or in the transportation and handling system. Yet the regulatory framework could not halt the approval of a product for the Canadian market on economic grounds, even if it spelled disaster. The crisis was eventually averted when Monsanto withdrew its application for approval for its GM wheat variety to be grown in Canada, but the case clearly illustrates why reliance upon this type of industry self-governance is risky.

At present, the Canadian industry with a stake in animal biotechnology is cautious about proceeding with animal biotechnology – derived products, because of nervousness about public opinion.[2] However, as products from cloned cattle and their progeny emerge into the food supply system in the United States, and inevitably enter into the Canadian market, it will be increasingly difficult to persuade Canadian industries to uphold the moratorium. It will also be difficult, if not impossible, for Canadian regulators to halt the movement of these products across the Canadian/US border, particularly cloned animals and their progeny, which are impossible to identify without a reliable traceability system in place (and no such system currently exists).[3]

The United States

In 2009, the United States Food and Drug Administration (USFDA) announced a regulatory framework for governing genetically modified animals, although at time of this writing no modified animals have yet been approved for market release into the food supply system. In contrast, cloned animals and their progeny are likely to be in the food supply system already, since the USFDA does not require any special regulatory approval for their market release (they are assessed in the same way as other animal products) and these products are not labelled.

The first official statement on animal biotechnology from the USFDA came with its publication in December 2006 of a draft risk assessment on animal cloning, which was adopted in January 2008 (USFDA 2006, 2008a). This much-anticipated document was the first government-produced risk assessment of animal cloning in the industrialized world and carried considerable weight, particularly as the United States houses the largest number of companies with an interest in animal cloning for the food industry. The risk assessment was accompanied by a 'Risk Management Plan' and 'Guidance for Industry' (USFDA 2008b, 2008c). The second official statement came in the form of a 'Draft Guidance on Regulation of Genetically Engineered Animals,' released for comment in September 2008 (USFDA 2008d) and published as a 'Final Guidance for Industry' in January 2009 (USFDA 2009).

The risk assessment for animal cloning is a purely scientific evaluation of the risks of animal cloning in agriculture. The USFDA relied on scientific data from two companies, and it concluded that food products derived from cloned animals and their progeny are safe for human consumption and that cloning poses no increased risk to animal health compared to other artificial reproductive technologies (ARTs) or natural breeding. Although it acknowledges that the frequency of adverse effects in cloned offspring increases compared to the progeny of ARTs, it believes that as the technology develops, this frequency will decline. The accompanying risk management plan defines risk management as a 'set of activities that integrate risk assessment results with other information to make decisions about the need for and method of risk reduction' (USFDA 2008b: 1). However, in the plan, the USFDA relies solely on the scientific risk assessment to draw risk management conclusions, stating that the ethical, moral, and religious issues, which it recognizes exist, are beyond its remit and 'unrelated' to its public health mission (ibid.: 5).

In 2001, the USFDA asked companies to implement a voluntary ban on introducing products from cloned animals and their progeny into the human or animal food supply system until the risk assessment could be completed. Publication of the USFDA's risk assessment, and the conclusion drawn in the risk management plan that products from cloned animals posed no new risks and required no additional or new regulatory controls, meant that this voluntary request could no longer be justified. It is now possible for products, such as milk and meat, derived from cloned animals and their progeny to enter the food chain in the United States under existing food safety regulations without any requirement for labelling or traceability.

Despite the USFDA position on animal cloning, a number of US companies implemented their own ban on meat and milk from cloned animals (but not from their progeny).[4] Smithfield Foods Inc., Kraft Foods Inc., Wal-Mart Stores Inc., and Tyson Foods Inc. all banned milk and meat from cloned animals from their products, based on consumer surveys that show consumers are uncomfortable with animal cloning on health, environmental, or ethical grounds (Zhang and Jargon 2008). However, a large number of companies did not institute such bans, making it impossible to know the degree to which products from cloned animals and their progeny were entering the food system.

In 2008, the USFDA released for public comment the 'Draft Guidance on Regulation of Genetically Engineered Animals' (2008d). The document was designed for industry feedback on the regulatory process and public education about the benefits of GE animals, which it lists as the health protection of animals, new sources of medicines, transplantation, less environmental impact, and healthier food (although no risks or concerns are mentioned) (USFDA 2008d). The USFDA proposed to regulate GE animals under the new animal drug provisions of the Federal Food, Drug and Cosmetic Act, so that the modified rDNA construct present in the GE animal would be treated as a drug.

Under this proposal, the USFDA must approve all GE animals before market release on a case-by-case basis. Depending on the level of risk, some GE animals may not require pre-market approval, for example, GE laboratory animals and the GloFish™. However, all food animals are expected to be approved before release to the market. New GE animals would also have to pass through an environmental review under the National Environmental Policy Act. The draft guidance triggered 28,000 public comments, which the USFDA responded to through a website (ibid.). The final guidance, entitled 'Regulation of Genetically Engineered Animals Containing Heritable Recombinant DNA Constructs' was released in January 2009 with little if any substantive changes to the draft guidance. Since this time, no GE animals have been approved for market release, although the agency is considering an application for release of GE salmon (modified for faster growth).

To date, the response to animal biotechnology in the United States has been almost exclusively at the regulatory level, and left to the USFDA. Consequently, it is fair to say that there has hardly been any officially sanctioned political debate at all. The USFDA is a science-based regulatory agency without the mandate or capacity to consider non-scientific issues or public concerns that are part of the political debate in many countries (Hartley and Skogstad 2005). The National Academy of Sciences (NAS) was critical of this narrow scientific remit in its published report on the scientific concerns of animal biotechnology in 2002. It recognized that animal biotechnology included a range of social, ethical, and religious concerns that could not be considered by the regulatory authorities due to their science-based remit (NAS 2002).

This problem of regulatory mandate was also flagged by the Pew Initiative on Food and Biotechnology (PIFB), established by the Pew Charitable Trusts in 2001, when the controversy around GM crops and foods was at a high point. The PIFB ran for six years and examined the social, ethical, political, and economic impacts of agriculture biotechnology, producing a number of reports and holding stakeholder forums. In 2004, the PIFB published a report on regulatory issues in animal biotechnology, and in 2005, it held a multi-stakeholder workshop, 'Exploring the Moral and Ethical Aspects of Genetically Engineered and Cloned Animals.'

The USFDA reviewed the Pew's findings and other recent public opinion research in a published paper in 2007 and concluded that public concerns were simply the result of lack of knowledge about the context and benefits of animal cloning and the scientific rigour in the regulatory framework (Rudenko and Matheson 2007). The USFDA saw a role for itself in ameliorating this knowledge deficit by providing accurate and unbiased scientific information through risk communication, arguing that 'putting cloning into the context of other ARTs, dispelling common misconceptions about clones, and explaining the utility of cloning in breeding programs could provide a factual basis from which the public could consider livestock cloning' (Rudenko and Matheson 2007: 203). However, the UK Food Standards Agency (2008) found that the public (at least in the UK) does not want to put cloning in the context of other ARTs. The Food Standards Agency research showed that the public found ARTs to be giving nature a 'helping hand,' whereas cloning was 'interfering with nature' and was seen as 'representing a step too far, one that involved crossing a line between natural and unnatural' (ibid.: 24–5). Perhaps more significantly, the Food Standards Agency research report clearly demonstrates that the USFDA's belief that public concerns are simply a result of lack of knowledge and education is seriously flawed, finding that as knowledge increases, concerns become more prevalent and strongly felt, particularly for women (ibid.: 28).

Europe

In line with the Canadian government, the governing bodies of the European Union have not made an explicit statement or provided guidance regarding GM animals. Existing GMO legislation is in effect until the public is notified otherwise. In 1992, the European Commission (the European Union's decision-making branch) asked the Group of Advisers to the European Commission on the Ethical Implications of Biotechnology to report on ethical aspects of GM animals, but there has been little activity since then. In contrast, there has been considerable activity on the issue of animal cloning in the European Union, culminating in the European Commission's 2010 proposal for a five-year ban on cloning for food and livestock production (European Commission 2010). Controversially, this decision cleared the way for products derived from the offspring and later descendants of cloned animals to enter the food system. At the time of writing, this decision had triggered a fierce debate across Europe.

In February 2007, prompted by the USFDA's risk assessment and the possible authorization of food products derived from cloned animals in the United States, the European Commission called on its scientific expert advisory panel, the European Food Safety Authority (EFSA), for a scientific opinion on food safety, animal health and welfare, and the environmental implications of animal clones, their progeny, and the food products derived from them. Released in July 2008, the EFSA scientific opinion was in line with that of the USFDA in that it did not consider the food safety risks from the products of animal cloning to be substantially different from those produced by other breeding methods (EFSA 2008). However, in contrast to the USFDA, the EFSA drew attention to the lack of adequate scientific research in the area (it noted the limited number of studies, small sample sizes, and absence of a uniform approach to data) and the uncertainty this brought to its conclusions. Perhaps more significantly, the EFSA raised concerns over animal welfare from a scientific perspective, noting that 'the health and welfare of a significant proportion of clones, mainly within the juvenile period for bovines and perinatal period for pigs, have been found to be adversely affected, often severely and with a fatal outcome' (ibid.: 2). This observation provides the scientific basis for the commission's ban on cloning for food production.

In June 2009 and again in September 2010, the EFSA published further statements on the scientific issues related to animal cloning at the request of the European Commission. Both times, the EFSA reviewed the latest scientific data and consulted with scientific experts, finally concluding that there was no evidence that would cause it to alter the findings and recommendations in its 2008 report (EFSA 2010).

At the same time the European Commission called on EFSA for its scientific opinion on animal cloning, it also called on its ethics advisers, the European Group on Ethics in Science and New Technologies to the European Commission (EGE) for an opinion on the ethical aspects of animal cloning for the food supply. The EGE based its opinion on expert hearings, public consultations, and roundtable discussions with a broad range of stakeholders. In January 2008, the EGE concluded that it could see no convincing arguments to support animal cloning in the food supply system (EGE 2008).

There is little commercial interest supporting animal cloning in the European Union (Farming UK 2009), but considerable political opposition, particularly from the European Parliament, much of it rooted in value-based social and ethical concerns. In September 2008, the European Parliament adopted a resolution calling on the commission to ban cloned animals and their offspring from the food supply system (European Parliament 2008). An overwhelming majority supported the resolution: 622 for, 32 against, with 25 abstaining (Weimer 2010).

There has been no survey of European public opinion on animal cloning through the Eurobarometer, Europe's primary mechanism for gauging public opinion. However, the European Commission funded a research project under its Framework 6 Program called Cloning in Public, conducted at the Danish Centre for Risk Assessment. This initiative facilitated a public debate about the ethical and societal consequences of animal cloning across the EU to test social acceptability, and it made policy recommendations on animal cloning (Gamborg et al. 2006). The research team found little public support for cloning in food production, a general lack of knowledge about cloning, and more concern for food safety, socio-economic impacts, and consumer choice than for animal welfare.

In October 2010, in response to pressure from the European Parliament, the European Commission announced the five-year temporary suspension of animal cloning for food production (including livestock production and food from clones) in the European Union (at the time of writing, the proposal had yet to be ratified by the EU). The European Commission (2010) states that the temporary ban is meant to allow time for the establishment of a traceability system for imports of reproductive materials for clones so that farmers and industry can set up a database for cloned animals. The commission clearly states that the ban is in response to animal welfare concerns and that there is no scientific evidence of food safety concerns. The decision falls in line with the EFSA's scientific advice, but does not reflect the EGE's ethical recommendations, the parliament's opposition, or public concerns, although it openly states that it considered the EGE's opinion in reaching its decision.

In reaching a decision, the European Commission had to consider its obligations under international law. The moratorium it issued on the commercial production and sale of genetically modified organisms in Europe in the late 1990s was challenged by the United States, Canada, and Argentina through the World Trade Organization (WTO). The three nations argued that the EU had contravened WTO agreements governing international trade that had the aim of removing barriers to trade and creating an open market. These legal agreements were negotiated and ratified by the majority of the world's trading nations. The General Agreement on Tariffs and Trade (GATT) and the Sanitary and Phytosanitary Agreement (SPS) apply to trade in food products and to animal biotechnology products. These agreements require that the EU conduct a risk assessment before issuing a moratorium, and that scientific evidence be provided demonstrating a threat to human health or to the environment: the EU needed to prove that GMOs were dangerous to humans or the environment on scientific grounds. The

EU argued that its moratorium was justified by the adoption of the precautionary principle in its assessment of these risks, which, it also argued, was a scientific principle. However, the WTO was not convinced, refusing to recognize the precautionary principle as a scientific principle, and ruled against the EU. Although the commission believes its decision to ban animal cloning is compatible with WTO rules (Ruitenberg 2010), it is unclear whether there will be a challenge in the future. Member states may still invoke the precautionary principle in market approval under pressure from public opposition at the national level (Weimer 2010).

Although the ban has generated criticism from the scientific community (for example, see ViaGen 2010), it has triggered a much broader and fierce debate across Europe because it allows for the entry of the progeny of clones into the food system. As animal welfare concerns cannot be attributed to the progeny of cloned animals, there is no scientific basis to restrict their entry into the food system or to label food products derived from them. These products will now fall under the Novel Foods Regulations (no. 258/97) and will be assessed on a product basis by EU member states. The process of cloning will not be considered in the assessment.

The EU countries were pressured by strong public concern to adopt a regime of mandatory labelling of foods containing genetically modified plant products, and will likely face the same pressure with respect to GE animal products and clones. However, a clone and the products derived from it cannot be distinguished from any other animal in biological terms, which means that labelling would have to be based on a strongly reliable (and very expensive) traceability system.

Reflections on Governance Frameworks

A brief look at the different governance frameworks that have evolved in response to animal biotechnology in Canada, the United States, and Europe provides some useful insights. The first of these insights is that it appears as though regulators in all these jurisdictions are constrained by a narrow scientific remit, which largely prohibits them from considering the full range of concerns held by stakeholders and the public. This narrow remit would not be as problematic if these broader issues were considered elsewhere in the political system, but in Canada and the United States, the official response to animal biotechnology comes from the regulatory level, not from any broader political level; therefore, the regulatory agency makes the decision about how products of animal biotechnology will be governed taking into account only scientific information.

This approach is problematic for a number of reasons. First of all, public policy making and regulatory development rely on democratic mechanisms for legitimacy, whether through mechanisms of participatory democracy, that is, consultation with members of the public and stakeholders, through representative democratic mechanisms, that is, parliamentary debate, or through functional democracy, that is, reliance on expert advice. When these democratic mechanisms are constrained by the requirement of considering scientific data alone, the democratic legitimacy of a policy or regulatory decision may be called into question (Hartley and Skogstad 2005). Second, reliance on a narrow scientific remit to govern a technology can be risky for industrial actors, who are necessarily sensitive to economic triggers in the marketplace. In the case of GM wheat, outlined earlier in this chapter, despite clear economic reasons for rejecting an application for the release of GM wheat due to fears about market collapse of the wheat industry, regulators could not reject the application. Values of any sort, even purely economic, cannot be considered in a rigorously science-based regulatory regime.

The second insight is that regulators and the regulatory frameworks for which they are responsible have been largely ineffective at picking up the issues associated with animal biotechnology. Consequently, the question needs to be carefully considered of where public values should be debated and incorporated into policy. The research represented in this book has identified a broad range of non-scientific issues directly related to the governance of animal biotechnology. A small number of these issues were identified in the CFIA and HC stakeholder consultations, and still more could have been identified with a broader range of stakeholders. However, it is clear that, despite the recognition that there are important value concerns, the CFIA and USFDA documents are unable to give them any weight in regulatory policy. Regulators are not given the legal remit to take the non-scientific issues into account. But we would argue that they probably are not well placed to do so – even if they had a broader remit. In contrast, following the functional model of democracy, the EU's EGE, comprising an independent set of experts, managed to capture the range of issues more effectively, although these issues were not reflected in the European Commission's proposal. Much research has been conducted on models of participatory democracy, where members of the public are given an opportunity to express their values on pertinent policy issues. But less is understood about how the value information derived from functional and participatory democratic models gets plugged into the policy-making framework. Policymakers still struggle with how to include public values in democratic institutions outside of the legislative process.

The third insight is that, in Europe and in North America, the issue of cloned animals has received priority attention by government and regulators over that of genetically engineered animals. We believe that the reason for this is fairly evident; it is due to the fact that the many ethical concerns surrounding human reproductive and therapeutic cloning have spilled over into the debate about non-human animal cloning, even though the arguments around cloning are quite different in the two cases.

In the case of human cloning, the concerns are largely articulated in terms of a concept of human dignity and respect for the person. Human reproductive cloning is generally viewed as problematic by ethicists because of the way in which it confuses the notion of personal identity and self-identity. How would a human clone view itself, and be viewed by others in terms of its relationship to the parent of which it is a clone? As a 'reproduction'? As a 'twin'? As a designer artefact? How does this affect a person's sense of selfhood and unique personhood? Human therapeutic cloning does not result in a new, cloned person, but only in the production of a cloned embryo from which tissue (e.g., stem cells) can be collected and used for therapy of the parent. The ethical issues raised in the minds of many in this case have to do with the status of the human embryo – its moral standing and its 'personhood.'

None of these ethical issues associated with human cloning, either reproductive or therapeutic, seem to be raised in the case of non-human animal cloning. Few ethicists believe that non-human animals (with the possible exception of some higher primates and other mammals) have levels of self-consciousness or moral awareness that would be impinged upon by cloning. The ethical issues around cloned animals centre primarily on the question of the impact on the health and welfare of the cloned animal or upon consumers of the cloned animal. These are matters that are in no way unique to the cloning technology itself.

Despite this very different moral perspective in the two cases, we believe that the ethical sensitivities around human cloning have transferred, in the concept of 'cloning' itself, to the case of non-human animal cloning, making it seem to regulators and industry stakeholders as the likely 'hot button' issue with consumers and the public. Genetically modified animals, on the other hand, are not seen as raising the same ethical sensitivities, probably because the ethical issues (still largely theoretical) around genetic modification of humans do not share a common concept (like 'cloning') that transfers to the case of non-human animals.

In addition, the cloning technology has also been the first to be brought to the market. It is the simpler of the technologies, and it could be argued that it is the easier to assess in terms of health and environmental risk.

Although it's too early to document the Canadian policy and regulatory position on animal biotechnology, we believe we can make some predictions about what will unfold, based on previous experiences with genetically modified crops, our focus group research, and personal communications with Canadian regulators. We think it is highly unlikely that Canada will adopt new legislation on animal cloning or GM animals. Instead, we suspect that Environment Canada, Health Canada, and the CFIA will rely upon the already adopted product/process concept that emerged in the regulation of GM crops to regulate the products of animal biotechnology on a product-by-product basis. This approach relies on the understanding that there is nothing new about animal biotechnology when compared to traditional methods of animal production in terms of the scientifically defined risks. A decision of this sort would mean that Canada would follow the line of thinking laid out in the USFDA risk assessment on animal cloning and would not require any new regulations for governing the products of animal biotechnology. If this is the case, these products will fall under the existing Environmental Protection Act, Novel Foods Regulations, Feeds Act and Regulations, Health of Animals Act, and Food and Drugs Act.

As in the US case, we suspect it will be the regulatory agencies that make the decision on whether a legislative debate is needed, and these regulatory agencies are constrained by a narrow science remit. We know from our focus group data and through personal communications that regulators recognize there are strong non-science issues associated with animal biotechnology and that these issues need to be debated, but, because of their science remit, they don't see these issues as their responsibility. Participants in the Regulators focus group also pointed out that the regulatory framework reflects the Canadian values of human and environmental health and that other Canadian values more specific to animal biotechnology should be dealt with through the parliamentary system. However, regulators are struggling with their mandate – both the CFIA and HC have held stakeholder consultations to unearth some of the non-science issues, even though they cannot consider them because of their science remit. Perhaps of more concern is that although regulators have an understanding that the non-science issues that are beyond their mandate may be significant in the governance of animal biotechnology, they seem to feel that it is not their responsibility to emphasize these issues to others in the governance framework. Despite the flaws in the stakeholder consultations held by HC and the CFIA, the data generated through these mechanisms is of interest; yet it cannot be considered in the development of a regulatory response to animal biotechnology, and it is not passed on to others for consideration. The CFIA speaks with authority on the issues when it says 'no new regulatory framework needed,' suggesting in a way that it has considered all the issues. We argue that either its mandate needs to be broadened so that it can consider all the issues or it needs to state clearly that it cannot consider the broader issues and can simply contribute to the decision on whether or not to have new broadening legislation by providing the scientific advice alone.

In terms of public access to information, we predict that industry confidentiality will trump the availability of public information about what products are coming down the pipeline, as it did in the case of GM plants. We think it is likely that GM and cloned animals will be appearing in net pens and on dinner plates without a great deal of public or consumer knowledge. Certainly, in the United States, it is impossible for consumers to know whether a meat or milk product is derived from cloned animals. It is even difficult for Canadian consumers to know with any certainty whether the meat and milk products they consume are derived from cloned animals, despite the ban of these products in Canada, because it is impossible for regulators to identify these products if they cross the border. In the case of GM animals, the USFDA has committed to holding its scientific advisory committee meetings in public before making decisions on GM animal applications under its review, but we suspect that much of the data that forms the basis for such a decision will be held back as confidential. We suggest that greater transparency is needed to help inform the public about what products are coming down the pipeline, or even earlier when products are being developed, so that the public can respond in effective ways.

With respect to the labelling of products from genetically modified animals, we suspect the CFIA will rely on the existing voluntary labelling regime currently in place for food products containing genetically modified plants. In the case of GM and cloned animals, however, even a voluntary regime is complicated by the extreme difficulty of identifying animal-biotechnology products without a system of traceability. We believe it unlikely that a system of traceability will be established to this end. The organic industry is likely to establish its own system of traceability to certify that organic products do not contain genetically modified or cloned animals or their progeny. The Organic Trade Association, which represents the $11 billion organic industry in North America, has already stated that national organic standards will not include products from cloned animals (OTA 2010). As is the case with genetically modified crops, consumers who wish to avoid products from cloned or genetically modified animals will have to pay a premium for organic goods due to the increased costs associated with protecting the integrity of the animals and their products in the face of an increasing prevalence of GM and cloned animals in the environment and the food chain.

These decisions around governance will mask public concerns about animal biotechnology unless a public backlash occurs – and it might. There is clearly stronger public sentiment against animal biotechnology than existed for plant biotechnology, as borne out in the focus groups and public surveys. It is likely that the first genetically modified animal to be approved will be the GM salmon, a fish that already generates considerable controversy on the west coast of Canada and North America. Government support of salmon farming in open water pens has been harshly criticized by vocal opposition groups with considerable public support. We suspect a genetically modified salmon is likely to generate considerable debate about environmental risks and values.

The issues we have identified in this book around the governance of animal biotechnology are not unique to this technology alone. The range of social and ethical issues posed by this technology is not unlike those raised by most new technologies. We have explored the ways in which various political systems in North America and Europe have difficulty in incorporating widespread social and ethical concerns about animal biotechnology into public policy, because of the narrow remit of government regulatory agencies to consider only science-based matters of health and environmental risk.

But this situation obtains for all new (and existing) technologies. It is strongly motivated and reinforced by the growing system of global trade, which views limitations upon technology that are not 'science based' as unfair barriers to trade. Most democratic societies, particularly in North America, do not have robust democratic institutions for taking the broader social concerns around technology development seriously and incorporating them into policy. The international climate of globalization is not encouraging the development of such institutions. Animal biotechnology, despite the widespread concerns about its ethical limits, will not likely be held within these limits. There is not great hope in the current political climate that other emerging technologies will fare differently.

NOTES

1 For a discussion of the ethical issues raised by reliance on the market as a regulatory tool, see Sagoff (1982).
2 Personal communications with industry stakeholders at a Genome Canada – funded workshop to prepare a Position Paper on Aquatic and Terrestrial Animal Genomics (see Plastow et al. for more details)
3 In the case of the GloFish™, once the USFDA had approved the GloFish™ for commercial sale, it began to cross the border and be sold in Canada despite the fact it had not be approved by Environment Canada under the New Substances Regulations. It took a year for Environment Canada to respond and halt the sale of the products (Marden et al., 2006).
4 It is almost impossible to guarantee that products do not include meat or milk from the progeny of cloned animals as there is no way to track or identify these animals. It is very unlikely that cloned animals themselves would be used in food products as these animals are so expensive to produce. Therefore, the ban on the use of clones does not hold a lot of weight.

References

Canadian Food Inspection Agency. 2004. Consultation on Animal Biotechnology. http://www.inspection.gc.ca/english/anima/biotech/2004/cabcbae.shtml.

European Commission. 2010. Commission favours temporary suspension of animal cloning for food production in the EU. IP/10/1349. Brussels: 19 October 2010. http://ec.europa.eu/dgs/health_consumer/docs/20101019_ip_ec_cloning_en.pdf.

European Food Safety Authority (EFSA). 2008. Scientific opinion of the scientific committee: Food safety, animal health and welfare and environmental impact of animals derived from cloning by Somatic Cell Nucleus Transfer (SCNT) and their offspring and products obtained from those animals. *EFSA Journal* 767: 1–49.

– 2009. Further advice on the implications of animal cloning (SCNT). http://www.efsa.europa.eu/en/scdocs/scdoc/319r.htm.

– 2010. Cloning. http://www.efsa.europa.eu/en/ahawtopics/topic/cloning.htm.

European Group on Ethics in Science and New Technologies to the European Commission (EGE). 2008. *Ethical aspects of animal cloning for food supply*. Opinion no. 23. Brussels: EGE.

European Parliament. 2008. *European Parliament resolution on the cloning of animals for food supply*. Text adopted Wednesday, 3 September. Brussels: European Parliament.

European Parliament Intergroup for the Welfare and Conservation of Animals. 2008. MEPs reject notion of cloning for food. http://www.animalwelfareintergroup.eu/.

Farming UK. 2009. European Commission guilty of ignoring its own legislation. http://www.farminguk.com.

Food Standards Agency. 2008. *Animal cloning and implications for the food chain: Findings of research among the general public*. London: Creative Research.

Gamborg, C., et al. 2006. *Regulating farm animal cloning: Recommendations from the project Cloning in Public*. Denmark: Danish Centre for Bioethics and Risk Assessment.

Hartley, S., and Skogstad, G. 2005. Regulating genetically modified crops and foods in Canada and the United Kingdom: Democratizing risk regulation. *Canadian Public Administration* 48(3): 305–27.

Kochhar, H.P.S., Adlakha-Hutcheon, G., and Evans, B.R. 2005. Regulatory considerations in biotechnology-derived animals in Canada. *Rev. Sci. Tech. Off. Int. Epiz.* 24: 117–25.

Kochhar, H.P.S., and Evans, B.R. 2007. Current status of regulating biotechnology-derived animals in Canada: Animal health and food safety considerations. *Theriogenology* 67(1): 188–97.

Marden, E., Longstaff, H., and Levy, E. 2006. The policy context and public consultation: A consideration of transgenic salmon. *Integrated Assessment Journal* 6(2): 73–97.

National Academy of Sciences (NAS), Committee on Defining Science-Based Concerns Associated with Products of Animal Biotechnology. 2002. *Animal biotechnology: Science-based concerns*. Washington, DC: National Academies Press.

Organic Trade Association (OTA). 2010. Statement on cloning. http://www.ota.com/pp/otaposition/cloning.html.

Plastow, G., et al. 2007. Aquatic and terrestrial animal genomics: A position paper prepared for Genome Canada.

Rudenko, L., and Matheson, J.C. 2007. The US FDA and animal cloning: Risk and regulatory approach. *Theriogenology* 67: 198–206.

Ruitenberg, R. 2010. European Union plans temporary ban on livestock cloning, cloned-food sales. *Bloomberg*, 19 October. http://www.bloomberg.com/news/2010–10–19/european-union-plans-temporary-ban-on-livestock-cloning-cloned-food-sales.html.

Sagoff, Mark. 1982. At the shrine of Our Lady of Fatima or why political questions are not all economic. *Arizona Law Review* 23: 1281–98.

United States Food and Drug Administration (USFDA). 2006. Animal cloning: A draft risk assessment. Center for Veterinary Medicine, Department of Health and Human Services.

– 2008a. Animal cloning: A risk assessment. Center for Veterinary Medicine, Department of Health and Human Services.

– 2008b. Animal cloning: Risk management plan for clones and their progeny. Center for Veterinary Medicine, Department of Health and Human Services.

– 2008c. Guidance for industry: Use of animal clones and clone progeny for human food and animal feed. Center for Veterinary Medicine, Department of Health and Human Services.

– 2008d. FDA releases draft guidance on regulation of genetically engineered animals. Center for Veterinary Medicine, USFDA. www.fda.gov/consumer/updates/ge_animals091808.html.

– 2009. Regulation of genetically engineered animals containing heritable recombinant DNA constructs. CVM GFI #187. Center for Veterinary Medicine, USFDA.

ViaGen. 2010. Euro Commission retreats from cloning technology. http://www.viagen.com/news/euro-commission-retreats-from-cloning-technology/.

Weimer, Maria. 2010. The regulatory challenge of animal cloning for food: The risks of risk regulation in the European Union. *European Journal of Risk Regulation* 1: 31–9

Zhang, J., and Jargon, J. 2008. US: Food companies pledge not to use clones. *Wall Street Journal*, 4 September.

Contributors

Conrad G. Brunk is emeritus professor of philosophy and past director of the Centre for Studies in Religion and Society at the University of Victoria, British Columbia. His areas of research and teaching include the ethical and religious aspects of environmental and health risk perception, and the communication and value aspects of science in public policy. Dr Brunk is a regular consultant to the Canadian government and international organizations on environmental and health risk management and biotechnology. He served as co-chair of the Royal Society of Canada Expert Panel on the Future of Food Biotechnology and from 2002 through 2004 as a member of the Canadian Biotechnology Advisory Committee. He is co-author with Lawrence Haworth and Brenda Lee of *Value Assumptions in Risk Assessment* (1991), co-editor with James O. Young of *The Ethics of Cultural Appropriation* (2009), and co-editor with Harold Coward of *Acceptable Genes? Religious Traditions and Genetically Modified Foods* (SUNY). Professor Brunk holds a PhD in philosophy from Northwestern University.

Harold Coward is founding director and emeritus fellow of the Centre for Studies in Religion and Society at the University of Victoria, British Columbia, and a fellow of the Royal Society of Canada. A specialist in Indian philosophy and religion, he is author of nineteen books, including *Pluralism in the World Religions*, *The Philosophy of the Grammarians*, *Yoga and Psychology*, and *Acceptable Genes? Religious Traditions and Genetically Modified Foods* (2009), co-edited with Conrad Brunk. His current research is focused on religious perspectives on death and dying in hospice palliative care.

Mickey Gjerris is an associate professor of bioethics at the Danish Centre for Bioethics and Risk Assessment at the University of Copenhagen, Denmark. Originally trained in theology, he moved into the world of applied ethics during his PhD studies and is today doing research within the areas of bioethics, climate change ethics, animal welfare, ethics of nature, nanotechnology, and ethics and the meaning of life. Working from a hermeneutically oriented phenomenological point of view, he seeks to integrate ethical thinking about new technologies within the broader context of human existence. He has written articles in all of above-mentioned fields, edited three books in Danish, and frequently tries to engage the public in ethical matters.

Sarah Hartley is an adjunct professor of political science at Simon Fraser University, British Columbia. Her research interests lie at the interface between science, values, and public policy, particularly in the formulation of public policy responses to plant and animal biotechnology and environmental risks. She established a GE^3LS (genomics-related environmental, ethical, economic, legal, and social issues) program at Genome British Columbia. In this role she was responsible for advising Genome BC and scientists on the social, ethical, and legal issues of genomics research, integrating social science and humanities research into genomics science, and building genomics-related capacity in the social sciences and humanities. Dr Hartley has a professional interest in broadening our understanding of expertise when solving research or policy problems.

Lyne Létourneau is assistant professor in the Department of Animal Sciences at Laval University, Quebec. She holds a doctorate in law from the University of Aberdeen (2000) as well as a master's (1993) and bachelor's (1988) in law from the University of Montreal. Her research interests concern the interface between regulation and ethics in animal protection and agricultural biotechnology. In addition to articles related to animal law and ethics, and ethical issues in the genetic engineering of animals, she is the author of *L'expérimentation animale: L'homme, l'éthique et la loi* (1994) and editor of *Bio-ingénierie et responsabilité sociale* (2006). Dr Létourneau served as a member of the Canadian Biotechnology Advisory Committee from 2002 to 2007.

Peter W.B. Phillips, an international political economist, is professor of public policy in the Johnson Shoyama Graduate School of Public Policy at the University of Saskatchewan. He undertakes research on governing transformative innovation, including biotechnology regulation and policy, innovation systems, intellectual property, supply chain management, and trade policy. He is co-lead and PI of the Genome Canada project on Value Addition through Genomics and GE^{3}LS (2009–14). His latest book, *Governing Transformative Technological Innovation: Who's in Charge?*, was published in 2007.

Nola M. Ries, MPA, LLM, is adjunct assistant professor in the School of Human and Social Development at the University of Victoria, British Columbia, and a research associate with the Health Law Institute at the University of Alberta. Her work focuses primarily on legal and regulatory issues in health law, including regulation of new biotechnologies. She has been involved in several Canadian research networks, including the Advanced Foods and Materials Network, where her research has addressed legal, ethical, and social issues related to genomics, food, and health products. She is a lecturer in health law and regularly presents at Canadian and international legal, medical, and scientific conferences.

Leslie C. Rodgers is a senior consultant with the Praxis Group™, a western Canada–based firm specializing in public consultation, social science research, and processes to inform policy development. Leslie has managed processes for engaging the public in topical issues such as genomics, biotechnology, and health care; corporate responsibility towards the environment and community; and social issues such as youth unemployment and graffiti. Praxis is an innovator in process design through the use of such tools as Delphi panels, web surveys, study circles, and STS™ (web-based stakeholder tracking systems) to 'translate theory into action.'

Lorraine Sheremeta is a lawyer and research officer at the National Research Council's National Institute for Nanotechnology and a research associate at the Health Law Institute at the Faculty of Law, University of Alberta. She is also an adviser to Alberta Innovates/ Technology Futures on nanotechnology R&D initiatives. Lori's current research interests include strategic risk communication and the legal and regulatory challenges raised by nanomaterials for the environment and human health. Her most recent publication is 'Small Is Different: A Science Perspective on the Regulatory Challenges of the Nanoscale' (2008), prepared for the Council of Canadian Academies. Lori is a long-term member of the Alberta Cancer Board Animal Ethics Committee and currently serves as a member on Health Canada's Science Advisory Board.

Paul B. Thompson holds the W.K. Kellogg Chair in Agricultural, Food and Community Ethics at Michigan State University. He has been an active teacher and researcher on the ethics of food production and consumption for over thirty years. Dr Thompson completed a PhD in philosophy at SUNY Stony Brook in 1980 and has held research and teaching posts at Texas A&M University and Purdue University. He is a two-time recipient of the American Agricultural Economics Association Award for Excellence in Communication and serves as a member on numerous advisory committees, including the US National Research Council's Advisory Committee on Biotechnology and Genome Canada's Science and Industry Advisory Committee. His current research focuses on the development of voluntary market standards for certification of animal welfare and the application of nanotechnology in agri-food.

Index